TIME-DELAY SYSTEMS

TIME-DELAY SYSTEMS

Editor

Dana Zelenko

scitus academics

Time-Delay Systems
Edited by **Dana Zelenko**

ISBN: 978-1-68117-657-4
Library of Congress Control Number: 2016946524

SCITUS Academics LLC,
616, Corporate Way, Suite 2, 4766,
Valley Cottage, NY 10989

www.scitusacademics.com

Notice

Printed in United States of America on Acid Free Paper

PREFACE

It is renowned that the delay information in systems that are matter-of-fact or perhaps processes should not be ignored in the modeling procedure, in combination with the increasing expectations of system dynamic performance. Time delays in the engineering community are a very popular phenomenon. In fact, many industrial processes involve after-effect phenomena in their internal dynamics. And actuators, sensors and field networks in feedback loops usually bring about inevitable delays. Time delays in dynamic systems are the inevitable result of the limited speed of transmission of information. Realistic models increasingly require that delays be included in order to understand, design, analyze and control real life systems effectively. Time delays are often observed in different technical systems, such as electrical, hydraulic and pneumatic networks, chemical processes, especially long transmission lines, robotics, etc. The question of controllability, observability, robustness, optimization, adaptive control, pole positioning and, in particular, stability and robustness stabilization for this particular system class has thus been one of the primary interests of many researchers and scientists over the last five decades. This book, Time-delay Systems, is concerned with the stability analysis and control problems of time delay systems; a research field with a history of at least six decades, which has been very active in the past two decades in particular.

TABLE OF CONTENTS

1

MARGINAL PROBABILISTIC MODELING OF THE DELAYS IN THE SENSORY DATA TRANSMISSION OF NETWORKED TELEROBOTS

Ana Gago-Benítez, Juan-Antonio Fernández-Madrigal and Ana Cruz-Martín *

Systems Engineering and Automation Department, University of Málaga, Campus Teatinos, Boulevard Luis Pasteur s/n, Málaga 29071, Spain

ABSTRACT

Networked telerobots are remotely controlled through general purpose networks and components, which are highly heterogeneous and exhibit stochastic response times; however their correct teleoperation requires a timely flow of information from sensors to remote stations. In order to guarantee these time requirements, a good on-line probabilistic estimation of the sensory transmission delays is needed. In many modern applications this estimation must be computationally highly efficient, e.g., when the system includes a web-based client interface. This paper studies marginal probability distributions that, under mild assumptions, can be a good approximation of the real distribution of the delays without using knowledge of their dynamics, are efficient to compute, and need minor modifications on the networked robot. Since sequences of delays exhibit strong non-linearities in these networked applications, to satisfy the*iid* hypothesis required by the marginal approach we apply a change detection method. The results reported here indicate that some parametrical models explain well many more real scenarios when using this change detection method, while some non-parametrical distributions have a very good rate of successful modeling in the case that non-linearity detection is not possible and that we split the total delay into its three basic terms: server, network and client times.

KEYWORDS

networked sensors; remote robot operation; stochastic time delays

1. INTRODUCTION

Telerobots, and, in general, platforms with remotely controlled sensors, are present in many advanced applications, like telecare robotics [1], telesurgery [2], underwater exploration [3], space operation [4], *etc.* In particular, mobile networked telerobots [5] are a class of mobile telerobots controllable over networks like the Internet, that are accessible to the general public through, for example, the World Wide Web. The main operation of such a mobile platform consists of receiving and executing motion commands that are issued from a remote user station, which in turn is displaying information acquired by the sensors of the robot, typically cameras, range sensor data, *etc.* (see Figure 1). This remote control task has time requirements that should be satisfied, because, e.g., the user cannot perform correctly robot navigation if the data are not received from the sensors at the client side with enough frequency.

Current Internet technology presents important problems to perform a hard real-time remote control of these robots due to its stochasticity, but not only the network is an issue in this regard: operating systems and application software, typical of these general-purpose applications, are problematic as well because they may inject unpredictable delays, in some occasions longer than the ones of the network, in a way that makes impossible to guarantee a timely information flow through the control loop in all situations. These are thus *soft* real-time systems [6], where only statistical requirements can be satisfied, for instance "the data from that sensor must arrive every 500 milliseconds with 90% of probability". A good probabilistic estimation of the delays in the transmission of the sensory data is consequently required to be able to predict the subsequent delays and thus make decisions that guarantee the probabilistic satisfaction of time requirements.

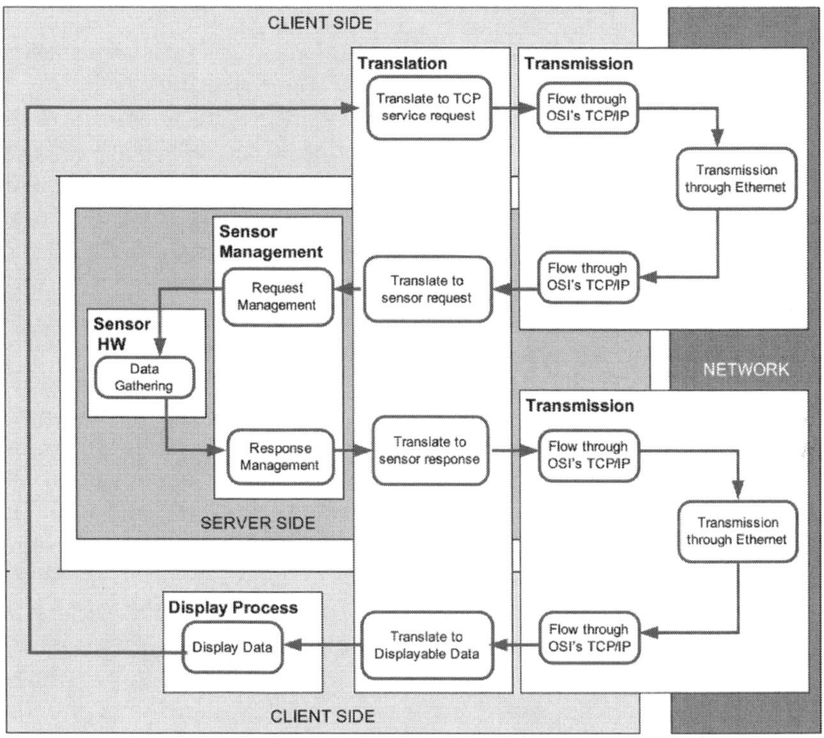

Figure 1. General scheme of a sensory loop in a networked telerobot controlled through the Internet, where both hardware and software components are non-deterministic in the real-time sense.

Different approaches, not necessarily robotic, may be found in literature to deal with stochastic delays of this kind: in the networking community they typically try to enhance the Quality of Service (QoS) of the network based on the modeling and control of the information flow [7,8], which usually is addressed by modifications in the network protocols or the hardware [9], or through more exotic approaches such as the injection of artificial delays in the transmission path in order to compensate the already existing ones [10]; in the automatic control area, different theoretical models have been proposed to cope with the full dynamics of remote or distributed systems dedicated to process control—which includes the dynamics of time delays—[11–13], and an important amount of research also is being developed on stochastic networked control systems (NCS), although it is common in these communities to consider the non-network components as deterministic and, usually, to know the dynamics of the plant to control. Finally, approaches to control the timing of data flow in multimedia applications also exist [8,14,15], but they do not need to cope with the strict time requirements of controlling robots. It is noteworthy that most solutions reported in the literature to the modeling and regulation of the delays occurring in networked systems only deal with the network part: typically the end-to-end delay, which includes only A/D conversion, packetization, network propagation, queuing and buffering [16], or the RTT (round-trip time), that refers to the delay existing between sending a package through the network and receiving its acknowledgment [17].

In this paper we are interested in analysing minimalistic mathematical models for all the time delays (not only network delays) found in the kind of applications described before—note that the networked telerobot applications can be generalized, under the perspective of the modeling and regulation of their delays, to any scenario where remote sensors send data to a station through a stochastic network and other non-deterministic software components. The approach presented here pursues to satisfy the following goals: the application requires the data to arrive before a particular time—at least under probabilistic constraints, as explained above—in order to be useful; minimal modifications can be done to the existing system; no previous knowledge is available about the dynamics of the delays; and only a low computational power is available. Taking into account all these constraints, we present a thorough statistical analysis of the time delays of the sensory flow for networked telerobots. We show that temporal sequences of delays gathered from such a system can be successfully modeled with simple statistical tools based on marginal probability distributions, especially when abrupt changes in the signal are appropriately detected—*i.e.*, quickly and with high sensitivity—since those abrupt changes can be then used for separating the sequence into segments that do not contain such non-linearities (in statistical terms, marginal probability distributions are correct models as long as there is temporal independence between values in the sequence). We analyse in the paper the scenarios where this approach is expected to work, and the general characteristics they show. Our marginal distribution approach has reduced computational complexity with respect to other methods, and maintains an appropriate level of accurateness. In particular, it provides statistical significance to the model, or, in other words, the models obtained can explain the data in a statistical sense.

Standard methods that are commonly applied to characterize this kind of sequences of random values work by representing the entire sequence by a single model that captures as accurately as possible all the dependences existing between the values, instead of separating the sequence into nearly independent segments as we propose here. The two most common approaches found in literature are time series and hidden Markov models (HMM), both with well-known computational costs [18] (they are also very often used off-line). On the one hand, time series come in several flavors, depending on their flexibility: ARMA models are $O(m^3T)$, where T is the length of the series and $m = p + q$ the sum of orders of the model—these orders are to be decided previously with some additional procedure— but they are unable to represent signals with abrupt changes or trends; when the series has trends we can use a more involved ARIMA model, which is $O(T^2)$[19], but it cannot deal with abrupt changes in the signal; when the signal is to be segmented due to the presence of such abrupt changes, real-time algorithms based on ARMA exist that are $O(m^3T^2)$ [20]; finally, more complex and specific time series algorithms can be found, but with even worse computational costs [21]. On the other hand, HMM deal naturally with signals that change abruptly, representing them as the output of a stochastic process that varies its (hidden) state probabilistically. Unfortunately, learning the parameters of an HMM usually requires a T that is longer than in the ARMA case, *i.e.*, to gather appreciably longer sequences of values; in addition, its complexity is $O(N^2T^2)$, where N is the number of states considered for the system. That number should be estimated previously with some other procedure.

In contrast to these standard approaches, our marginal distribution method is $O(T)$, since the estimation of the distribution parameters and the change detection algorithm are $O(T)$ at most: on the one hand, in the worst case when we use a non-linear optimization algorithm to find the parameters of the distribution—which is avoided if we employ maximum likelihood estimation, as it is the case in many of our models, such as the lognormal [22]— the number of optimization steps can be bounded to a constant in all practical cases; on the other hand, the change detection algorithm that we present in Section 4.1 consists in the application of a mathematical expression, which is simply $O(1)$. Also, although our approach is specifically devised to detect and rule out abrupt signal changes, it produces improvements in the modeling of the signal when the delays have smooth (non abrupt) trend changes, since the procedure partitions the sequence of delays into shorter segments where the trend has a relatively lesser significance.

The first result reported in the paper is a statistical analysis, based on a diverse dataset gathered from very different real scenarios, that shows how most typical situations have delays with negligible autocorrelation as long as abrupt changes are detected, and produce long enough near-stationary regimes in that case, thus enabling the assumption of independence and consequently the application of our marginal model. After that, we show how our marginal approach applies to modeling the total delay in the sensory loop of many real scenarios, not only the network delay. Finally, our study proceeds to establish that some heavy-tailed marginal distributions such as the lognormal, along with our confidence interval method for detecting abrupt signal changes, highly improve the modeling of the delays, without requiring any previous knowledge about the dynamics of the system (elsewhere we have reported more in depth results for other heavy-tailed distributions when used with a different, more computationally involved change detection algorithm [23], that in this paper is used only as a near-optimal bound of the results that can be obtained with our marginal approach). The confidence interval method introduced in this work for detecting non-linearities does not overload the control loop—it is intended to be executed at the client side—and thus it is a promising start for further practical applications that use the proposed models in platforms with low computational capacity. Apart from these results with parametrical distributions, we also report results showing that some non-parametrical models are able to model correctly the delays of most scenarios of interest when they are just split into three basic components, something that is easy to do in many situations. This is practical in those cases where we cannot detect and process abrupt changes in the delay signal.

The document is structured as follows: Section 2 describes our experimental settings and justifies in which cases the marginal distribution is a suitable approximation of the stochastic behaviour of the delays; Section 3 proposes theoretical models of the time delays and evaluates their goodness-of-fit when confronted with the unprocessed experimental data; Section 4 introduces the confidence interval method to detect outliers, bursts and regime changes, and provides the results of fitting and hypothesis testing on the same scenarios, complementing them with the illustration of near-optimal results that would be obtained if the change detection algorithm is allowed to be more computationally involved. Finally, some conclusions and future work are outlined.

2. OVERVIEW OF THE EXPERIMENTAL SETTING

We have set up a number of scenarios (listed in Table 1) for covering the diversity of combinations of non-deterministic components that can be found in many networked telerobot systems, thus obtaining a very complete idea of the possible behaviours of the delays in the transmission of sensory data. Some of these scenarios have been previously used in other works (e.g., [23]), but only with one particular model of the delays, and using different, more computationally involved change detection methods. In this paper the number of scenarios is much larger and we use them for analysing many models simultaneously; we also split some of them into their fundamental parts, and provide, in general, a both wider and deeper analysis of the problem.

One of the scenarios (#13) does not belong to the scope of the problem discussed in this paper: its stochasticity in the delays is practically null. It has been included here for analysing our algorithms in the limit case that delays are almost deterministic instead of stochastic. In reality it could correspond to a very small microrobot with a 8-bit microcontroller on-board, connected to a external PC that requests sensory data through an USB cable (see description below).

Table 1. Summary of the experimental scenarios that have been set up for modeling the delay in sensory data transmission from our networked telerobots (more details on the elements of this table can be consulted in the main text).

Scenario	Robot	Sensor	Resolution	Location	Client Computer	Server Computer	Network
#1	SANCHO	Laser	181 data points	Same building	A	a	1
#2	SANCHO	Webcam	20% full B&W	Same city	B	a	2
#3	SANCHO	Webcam	40% full B&W	Same city	B	a	2
#4	SANCHO	Webcam	30% full B&W	Same city	B	b	3
#5	SANCHO	Webcam	100% full color	Same city	B	b	3
#6	SANCHO	Webcam	10% full B&W	Same building	A	b	4
#7	SANCHO	Webcam	50% full B&W	Same building	A	b	4
#8	SANCHO	Webcam	100% full B&W	Same building	A	b	4

Scenario	Robot	Sensor	Resolution	Location	Client Computer	Server Computer	Network
#9	SURVEYOR	Webcam	5.86% full color	Same building	C	c	5
#10	GIRAFF	Odometry	9323 bytes	Same building	D	d	1 + 4
#11	PHP simulated odometry only	Null (echo)	20 bytes	Ituzaingó (Argentina) & Málaga (Spain)	E	e	4 + 6 + 7
#12	C++ mobile robot simulator	Odometry	80 bytes	Same city	F	e *	2+4
#13	Arduino	Analog input	27 bytes	Same desk (p2p)	G	f	8
#14	PHP simulated complete robot	Null (echo)	4 bytes	Same building	H	h	4
#15	PHP simulated complete robot	Simulated camera	100% RGB	Same computer	H	g	No net
#16	PHP simulated complete robot	Simulated laser	2048 data points	Same building	H	h	4
#17	PHP simulated complete robot	Simulated camera	100% RGB	Same computer	H	h	4
#16	PHP simulated complete robot	Simulated laser	2048 data points	Same building	H	h	4
#17	PHP simulated complete robot	Simulated camera	100% RGB	Same computer	H	h	4
#18	PHP simulated complete robot	Simulated camera	100% RGB	Plymouth (UK) & Málaga (Spain)	I	e	4 + 6 + 7

Legend: A: Pentium M@1.8 GHz, 1 GB RAM, LINUX, FireFox. B: Intel Core Duo T7200@2 GHz, 2 GB RAM, LINUX, FireFox. C: Android smartphone Galaxy SII. D: Desktop PC Intel R CoreTM i5 3330@3 GHz, 8 GB RAM, Windows XP, teleoperation graphical interface of [24]. E: Laptop Intel core 2 duo T6400@2 GHz, 2 GB RAM, Windows XP. F: Laptop Intel core 2 duo T7200@2 GHz, 2 GB RAM, Linux. G: Desktop PC Intel core i7 960@3.20 GHz, 12 MB RAM, Linux. H: Desktop PC Intel core i7 940@2.93 GHz, 12 GB RAM, Linux. I: Laptop Intel Core 2 Duo P9600@2.53 GHz, 4 GB RAM, Linux. a: Intel Pentium M@2 GHz, 1 GB RAM, LINUX. b: Intel Pentium IV@3.2 GHz, 1 GB RAM, WINDOWS. c: Surveyor, Android. d: Mini-PC, Intel core i3 M@2.10 GHz, 3 GB RAM, Windows Embedded Standard. e: Desktop PC Intel core i7 950@3.07 GHz, 16 GB RAM, Linux. f: Arduino board with Atmel Atmega 328P@16 MHz, 2 KB RAM, no OS. g: Desktop PC Intel core i7 940@2.93 GHz, 12 GB RAM, Linux. h: Mini PC Intel core i5 650@3.20 GHz, 4 GB RAM, Windows 7 Enterprise N. 1: WiFi 802.11b/g, local University provider in the same building as the robot. 2: WiFi 802.11b/g, ISP provider located in Madrid (about 600 km from our lab). 3: RTC 56 Kbps, ISP provider located in Madrid (about 600 km from our lab). 4: Twisted-pair 1GBs Cable. 5: WiFi *ad-hoc*. 6: Submarine data cables [25]. 7: WiFi 802.11b/g ISP provider in Ituzaingó, Argentina. 8: USB 2.0 point-to-point cable.

*: robot simulation running in a virtual machine with Windows XP and the BABEL development system [26].

The following are the non-deterministic components that set up our scenarios:

- *Remote sensors*: We have used a mobile service robot called SANCHO [27] with: (i) a USB color camera sensor (webcam) with 48-bit color depth and 640 × 480 maximum resolution (it also captures B&W pictures) and (ii) a SICK LMS-200 laser scanner providing 360, 180 and 90 range data, connected through RS-422. The second robot we have employed in our experiments is the

Surveyor microrobot [28], which has a CMOS image sensor with 1.3 megapixels requesting 320 × 240 images (5.6% of the full color resolution). The Surveyor robot sends compressed images (JPEG), which randomly alter the actual density of the transmitted data. In the experiments we have considered only this single fixed resolution in order to carry out our analysis consistently. The third robot in our experiments is the Giraff [26], a robotic assistant for the elderly from which we have used the odometry sensors and the camera. We have also included in the experiments some simulated robots: (i) a simple simulation of the odometry of a differential wheeled robot written in PHP (for being executed by the same web server that is receiving the sensor requests); (ii) a complete robot simulation, also in PHP, that includes cameras (only noisy data is served, but with different resolutions); and (iii) a much more realistic simulation written in C++ that have been used in the last decade for a number of robotic experiments in our lab (it resembles the SANCHO robot, including laser scanners, odometry, reactive navigation, *etc.*). Finally, we have used a non-robotic device for simulating microrobots smaller than the Surveyor: an Arduino board [29] that only serves readings from one of the analog input channels of its microcontroller. This system provides much more deterministic delays than all the others, therefore it is especially difficult to model by our stochastic approach.

- *Servers for the sensory system:* The camera sensor of SANCHO is plugged-in to an Intel Pentium IV@3.2 GHz laptop with 1 GB RAM. The laser sensor is connected to a second on-board laptop (Intel Pentium M@2 GHz with 1 GB RAM). Both laptops are connected to the local network through Ethernet, with only twisted pair segments at 1 Gbps (the robot is stopped), and run MS Windows XP and also a robotic software mini-architecture built on the CORBA middleware [30]. The Surveyor robot uses a proprietary firmware that allows a remote client to request specific commands for getting images and other actions. The camera of the Giraff robot and its motion/odometry system are connected to the onboard mini-PC (Intel core i-3-M@2.10 GHz, 3GB RAM, Windows Embedded Standard), which is connected through WiFi to the router of the lab where these experiments have been carried out.

- *Web servers*: For SANCHO, an Apache web server is located in the same sub-network as the robot. It runs GNU/Linux (Ubuntu), and serves a PHP [31] page that connects through TCP/IP sockets to the robot middleware. The Surveyor, in contrast, includes an onboard *ad-hoc* web server that processes remote requests directly. The

Giraff robot does not use a web server since it is not managed through a web-based interface, but an *ad-hoc* one [24]. The PHP simulated robots are run by the same web server that receives the sensor requests (on Linux). The C++ simulated robot run on the same machine as this web server, but within a virtual machine with Windows XP. The Arduino system does not have any web server.

- *Web clients*: The client web browser for SANCHO has been Firefox on Linux, where a piece of Asynchronous JavaScript and XML (AJAX) code [31] continuously requests data from the sensors and measures the time delays. Two different laptops have been tested for running this web client: an Intel Core Duo T7200@2 GHz with 2 GB RAM (running Mandriva Linux) and a Pentium M@1.8 GHz with 1 GB RAM (running Ubuntu Linux). Both can use WiFi 802.11b/g connections to access the network: the former through an ISP provider located in Madrid (at about 600 km from our lab); the latter through a local University provider in the same building as the robot. In addition, we have included in some experiments a 56 kbps narrowband segment by using an old phone modem. For the Surveyor robot, a specific Android client application connects to the robot onboard web server, using the robot control protocol and a WiFi *ad-hoc*(single-hop) connection, and retrieving camera images. The Giraff robot is managed through an *ad-hoc* teleoperation graphical interface written in C++ [24], that is run in an external desktop PC (Intel R CoreTM i5 3330@3 GHz, 8 GB RAM, Windows XP). The PHP simulated robots have a simple web interface that has been run on three different computers (Laptop intel core 2 duo T6400@2 GHz, 2 GB RAM, Windows XP; Desktop PC Intel Core i7-940@2.93 GHz, 12 GB RAM, Linux; Laptop Intel Core 2 Duo P9600@2.53 GHz, 4 GB RAM, Linux) with Firefox and Chrome browsers. The C++ simulated robot has the same web interface but run in a different computer (Laptop Intel core 2 duo T7200@2 GHz, 2 GB RAM, Linux). Finally, the Arduino system has no interface: direct sensor requests are conducted through the USB connection from a testing program, also written in C++, that runs on a Desktop PC Intel core i7 960@3.20 GHz, 12 MB RAM, Linux.

As we have already reported elsewhere [23], since our approach is aimed at not using any knowledge about the plant that is producing the delays, its applicability cannot be demonstrated analytically for every existing combination of network, computer hardware and software, but notice that the components enumerated above are representative of a wide range of practical, non hard real-time applications: concerning the network, some of our setups include a number of hops, since they have been carried out from geographically distant places—including transatlantic submarine Internet cables—while others have been performed just with one network segment in the system or even a simple point to point connection; the hardware of the robots is obviously fixed, but both are very different and in particular have different computing power, based on a microcontrollers, standard laptops and mini-PCs; in addition we have used common robotic sensors, which include cameras to justify the applicability of our approach to a wide diversity of scenarios, because they allow us to largely vary the amount of data transmitted; finally, the software used in all the experiments is for standard general-purpose machines (except for the Surveyor firmware and the Arduino board), and we have no restriction in working with any reasonable number of applications, not related to our experiments, while they were executed, including modules in charge of other tasks in the robot, office software, Internet navigation in the case of the client controller (with different web browsers and OSes), and other applications in the case of the web server.

Moreover, the state-of-the-art literature on networked telerobots shows that these components are the most common in this kind of robots (see, for instance, [32,33]). In [34] you can consult a list of usual telerobots and camera sensors. Regarding the communication links, [32] presents a survey of wired and wireless connections (they refer to 2006, though the technologies presented are still valid). In Table 2 we summarise two examples of this kind of robots and their components.

In stochastic systems like the ones described before, the total delays of the sensory loop (*i.e.*, network + robot + client and server software) may vary over time due to different causes: transmissions from other sensors, network congestion, non-real-time operating systems in the loop, unpredictable behaviour of the client and the robot software applications, *etc.* Note however that it is unlikely that the delay produced when requesting data from a sensor depends on the values of previous delays measured for the same sensor, provided both that the sensor is not requested again until the previous requests end, and that the system has no explicit memory of these delays. Therefore, our hypothesis is that the dependences appearing in a sequence of delays will be strongly determined by changes in the underlying stochastic state of the components of the system, out of our reach.

Table 2. Two networked telerobots (one non-mobile and the other mobile) as examples of the usual kind of components that are found in these systems.

Project	Physical Structure	Sensors	Communications	Software	Interface
The Mercury Project	IBM SR5427 robotic arm by Sankyo	CCD Camera gray 192 × 165	Internet, Ethernet card, serial line	Server A: Sun SPARCServer 1000, SunOS 5.3 Server C: PC, MS-DOS	Camera image X,Y,Z coordinates Robot Movement (low-level commands)
Xavier	B24 mobile base by Real World Interface	Bump panels Wheel Encoders 24 sonar ring 30° fov laser Color camera + pan-tilt Speaker Speech-to-text card	Wireless Ethernet Thin-wire Ethernet	Two PC computers and a 486 laptop, all of them running Linux Sparc5 workstation with Netscape webserver	Camera image Zoomable map Robot Movement (high-level commands)

Figure 2 shows the sequences of total delays that we have measured in the sensory loop for each of the scenarios defined in Table 1, versus the time when each delay value was measured; it is easy to observe abrupt changes of regimes, outliers and bursts in those delay signals. Also, some of them exhibit smooth changes in their average, *i.e.*, their trends vary. In that figure we have drawn thick lines for separating portions of the scenarios where we cannot detect visually relevant abrupt changes or other non-linearities (*i.e.*, these portions are near-stationary, also called "regimes"), and marked some clear bursts. The boxplots of Figure 3 confirm the wide range of time delays measured and the pronounced skewness of their marginal distributions.

As explained in the introduction, it is a suitable approximation to assume statistical independence of sequential delay values in the stationary parts of the signal, which will allow us to apply the basic tools of the next sections and, in general, work with marginal distributions as suitable models of the data. To confirm this, we have analysed the correlation coefficients in the acquired data: the delays would be dependent if the autocorrelogram function (ACF) goes above the corresponding confidence limits; otherwise, our independence hypothesis cannot be rejected [35]. Figure 4 shows the ACFs of the scenarios, where it is clear the strong dependence of many of them when non-stationarity is not detected and handled, and also the ACF of the visually purged scenarios (eliminating the visually detected bursts and regimes marked in Figure 2). We can see how the ACF stays closer to zero in the latter, and it remains below the confidence limits in most cases. This supports our assumption of considering *iid* sequences of delays (independent and identically distributed) as long as they are separated in regimes and we can also detect bursts and isolated outliers. Of course, smooth variations on the underlying distribution—smooth trend changes—are still possible, but the results obtained with our abrupt detection method also applies in many cases that exhibit such trending.

For this article, we have classified the scenarios of Table 1 into three different categories, depending on the kind and combination of non-linearities that can be observed in Figure 2. This classification is subjective and a priori, but it is interesting because it reflects some important features of the signal. We have defined, by simple visual inspection of the data, three classes:

- Some of the scenarios do not exhibit clear abrupt changes, but do show a smooth fluctuating trend instead. We set these as class A (scenarios #1, #9, #3reg1—corresponding to the left part of scenario #3, Figure 2, which we have visually split into two regimes #3Reg1 and #3Reg2—, #10, #12, #15, #16, #17).

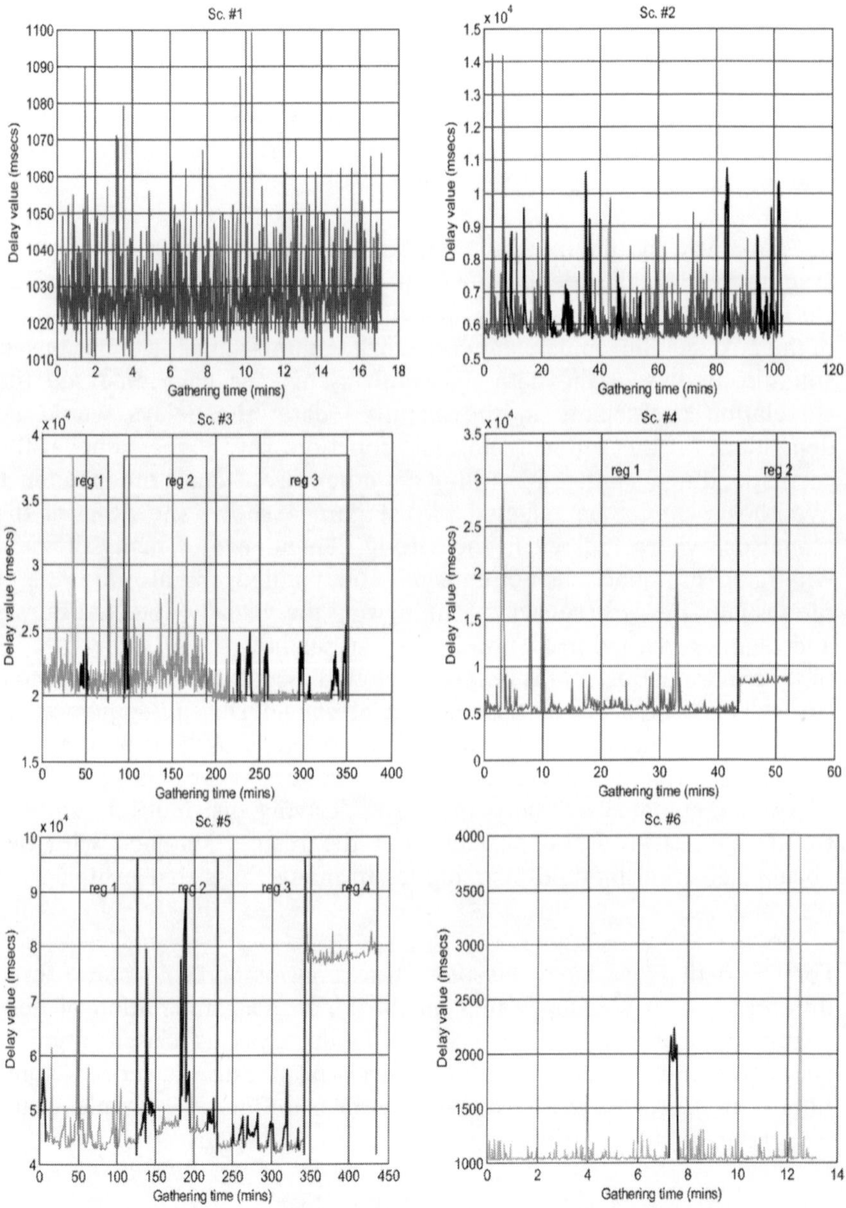

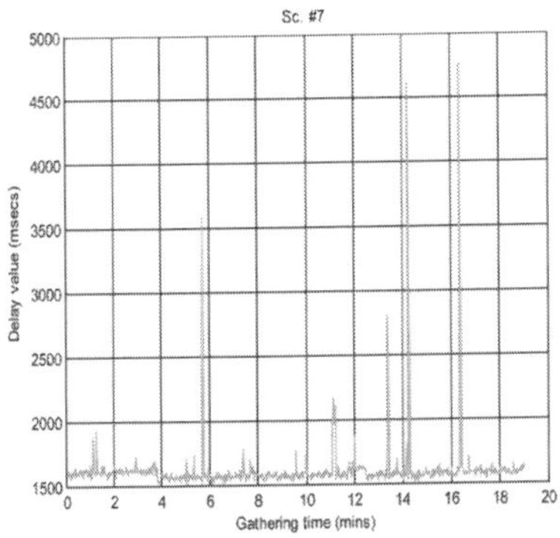
Sc. #7
Delay value (msecs)
Gathering time (mins)
5000
4500
4000
3500
3000
2500
2000
1500
0 2 4 6 8 10 12 14 16 18 20

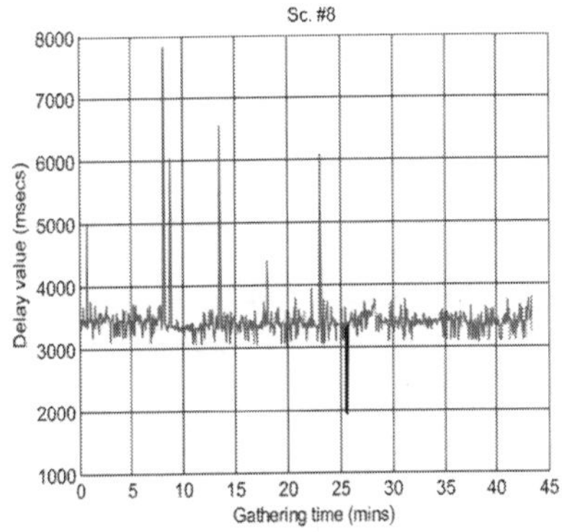
Sc. #8
Delay value (msecs)
Gathering time (mins)
8000
7000
6000
5000
4000
3000
2000
1000
0 5 10 15 20 25 30 35 40 45

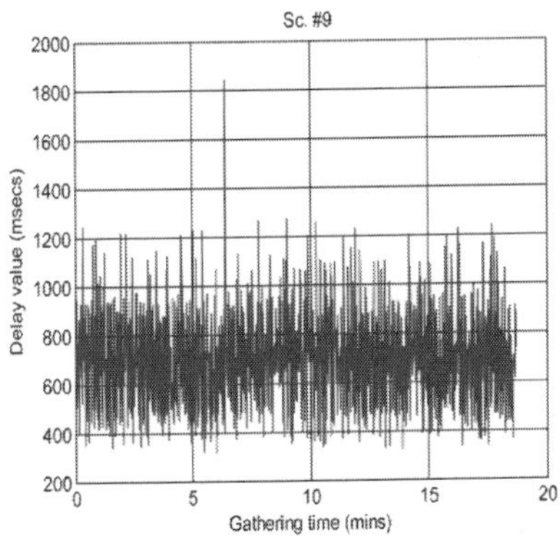
Sc. #9
Delay value (msecs)
Gathering time (mins)
2000
1800
1600
1400
1200
1000
800
600
400
200
0 5 10 15 20

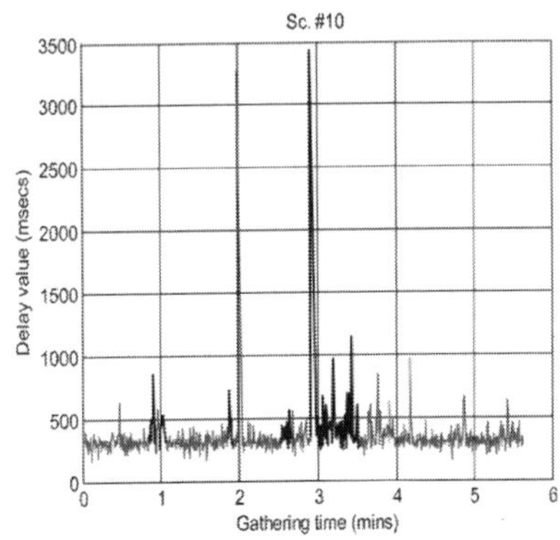
Sc. #10
Delay value (msecs)
Gathering time (mins)
3500
3000
2500
2000
1500
1000
500
0
0 1 2 3 4 5 6

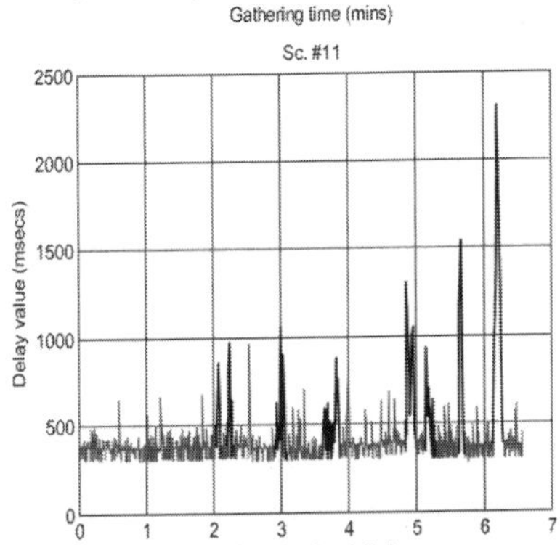
Sc. #11
Delay value (msecs)
Gathering time (mins)
2500
2000
1500
1000
500
0
0 1 2 3 4 5 6 7

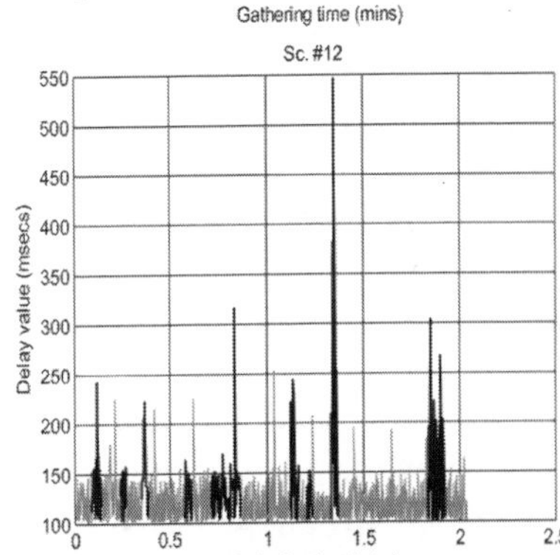
Sc. #12
Delay value (msecs)
Gathering time (mins)
550
500
450
400
350
300
250
200
150
100
0 0.5 1 1.5 2 2.5

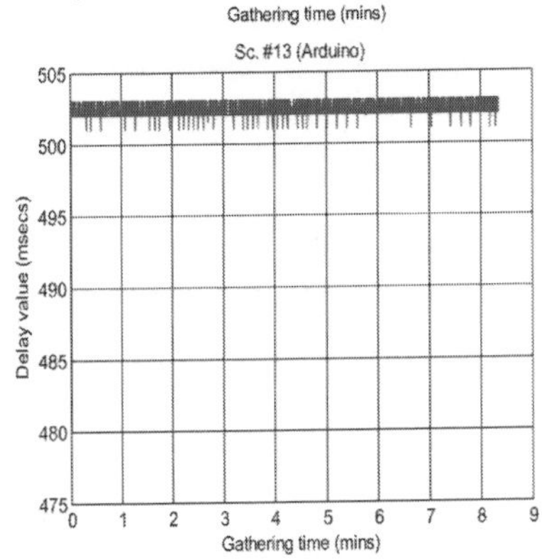
Sc. #13 (Arduino)
Delay value (msecs)
Gathering time (mins)
505
500
495
490
485
480
475
0 1 2 3 4 5 6 7 8 9

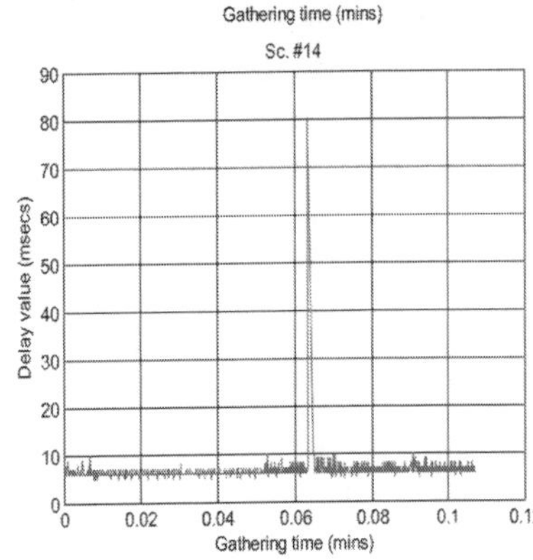
Sc. #14
Delay value (msecs)
Gathering time (mins)
90
80
70
60
50
40
30
20
10
0
0 0.02 0.04 0.06 0.08 0.1 0.12

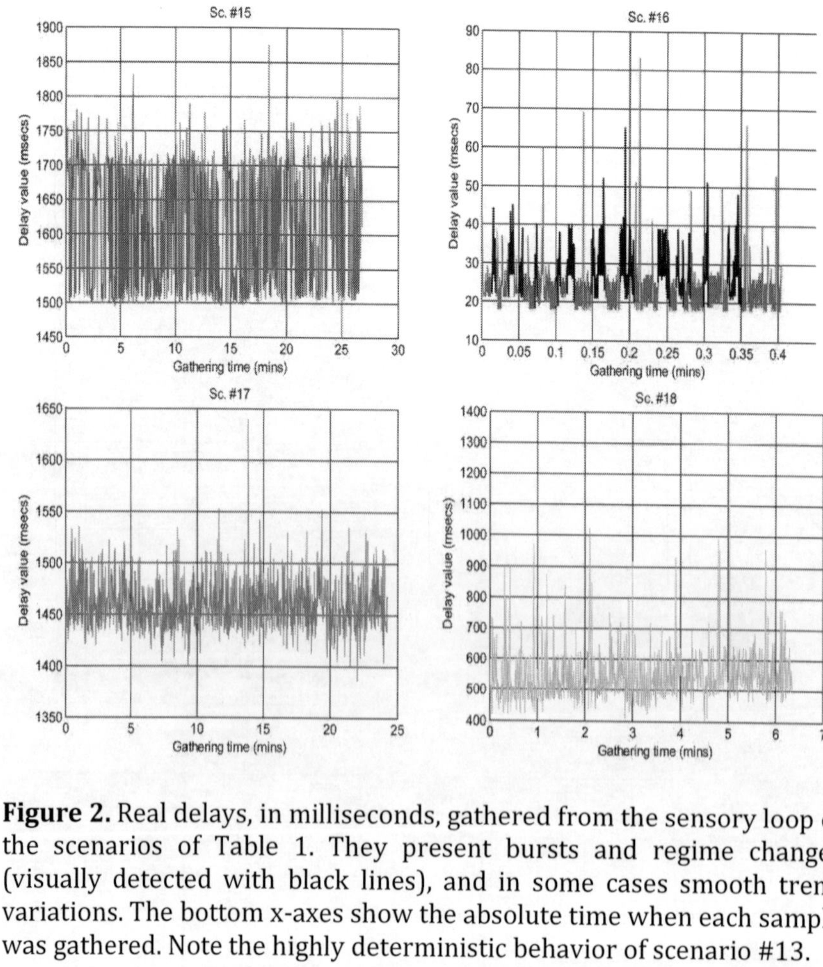

Figure 2. Real delays, in milliseconds, gathered from the sensory loop of the scenarios of Table 1. They present bursts and regime changes (visually detected with black lines), and in some cases smooth trend variations. The bottom x-axes show the absolute time when each sample was gathered. Note the highly deterministic behavior of scenario #13.

- Another broad group of scenarios do not show any continuous, smooth changes in their trend, and may or not have abrupt changes of regimes or bursts. They are class B, which can be further sub-divided into sub-class B1, on the one hand, for the scenarios that do not have those abrupt changes or that have only one dominant regime in the signal (scenarios #2 and #18), and sub-class B2, on the other hand, for scenarios #3reg2, #4, #5, #6 and #7.

- Finally, we have consider a special class C for scenarios that have neither abrupt changes of regime or bursts nor relevant fluctuation of their trends—thus, in principle they could be in class B—but that present outliers that are much smaller than the majority of its sequence of delays, something that does not occur in other classes and that we believe can have consequences in its modeling with many of the distributions explained in Section 3.1—due to the fact that their histograms do not follow the typical shape of a long-tailed distribution on its left-side portion. Scenarios #8, #11, #13 and #14 can be assigned to this class.

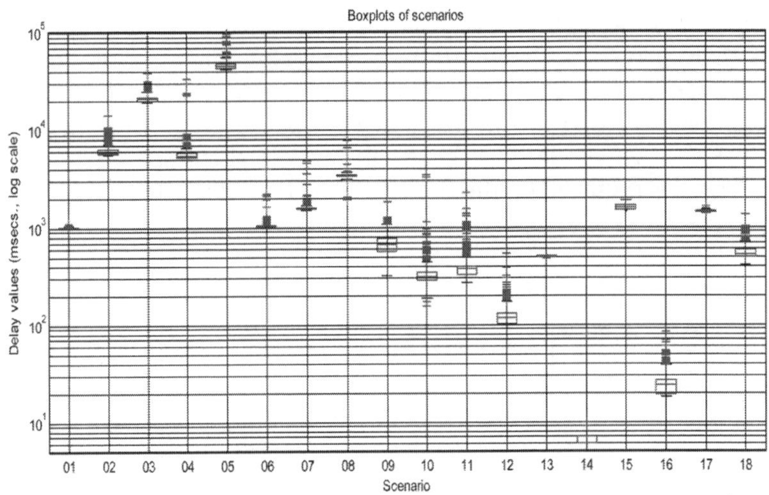

Figure 3. Boxplots of the scenarios of Figure 2. The y-axis is shown in logarithmic scale to appreciate better the amount of delay values (red markers) that are above the principal quartiles (blue boxes), which demonstrates the skewness of the marginal distributions.

One purpose of this preliminary visual classification is to illustrate better (*i.e.*, more intuitively) further experimental results. In principle, those results should go as follows: class B1 should produce good models due to its lack of abrupt changes and trend variations, and the same should occur for class B2 as long as abrupt changes are detected, since that can be enough to reduce the effects of the trends; class A is in a similar situation as B2: it can produce good models as long as the detection method reduces the mentioned effects of the changing trends; finally, it is not clear how class C will be modeled by our approach, since in principle it should have good models for its lack of abrupt changes and trends, but the presence of the special outlier makes that uncertain.

3. MODELING UNPROCESSED DELAYS

In this section we present how the original scenarios, without filtering out their non-linearities, fit directly with parametrical, non-parametrical and mixed marginal distribution models, which is assessed by two goodness-of-fit tests. This illustrates the basic methods employed (summarised in Section 3.1) and states their limitations in that initial situation. After showing these modeling results in Section 3.2, Section 3.3 is devoted to improve them—still without filtering the data—through a simple splitting of each delay into three different additive parts that correspond to the three general components that constitute our networked telerobot: the delay in the client side, in the network, and in the server (robot). We analyse whether that splitting, which involves a negligible overhead in the algorithms and in the modifications required on the system, leads to better results.

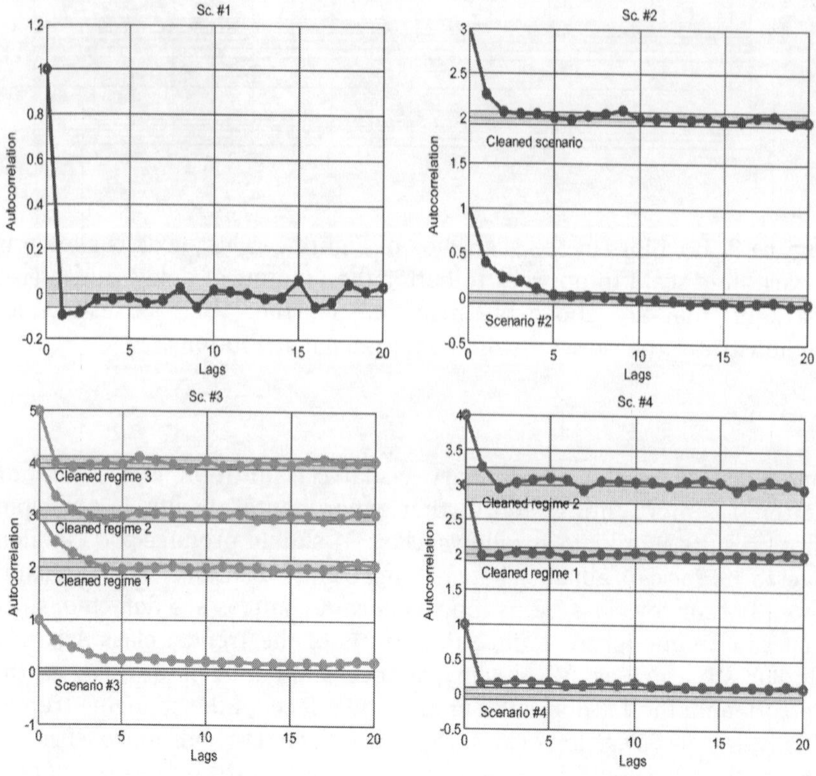

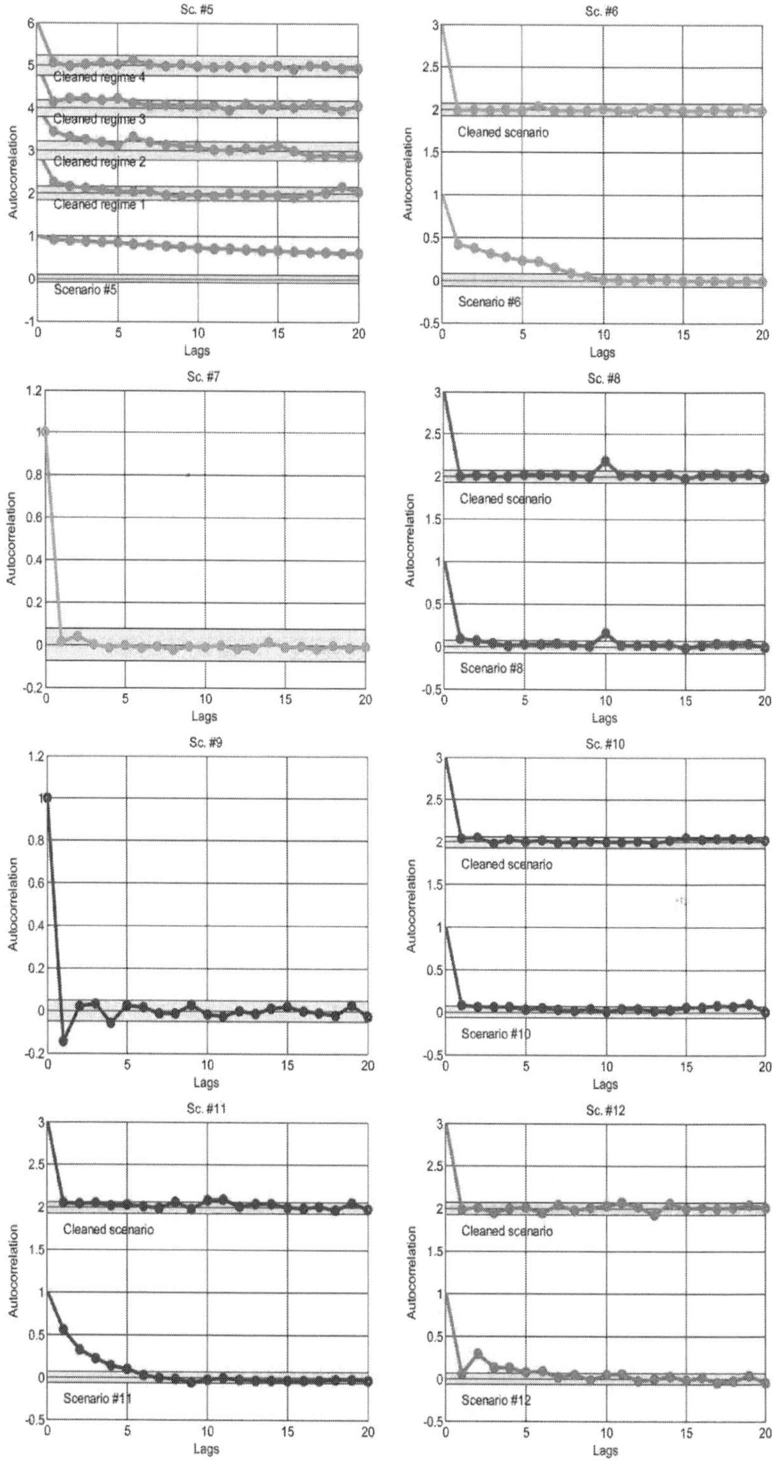
Sc. #5
Cleaned regime 4
Cleaned regime 3
Cleaned regime 2
Cleaned regime 1
Scenario #5
Sc. #6
Cleaned scenario
Scenario #6
Sc. #7
Sc. #8
Cleaned scenario
Scenario #8
Sc. #9
Sc. #10
Cleaned scenario
Scenario #10
Sc. #11
Cleaned scenario
Scenario #11
Sc. #12
Cleaned scenario
Scenario #12
Autocorrelation
Lags

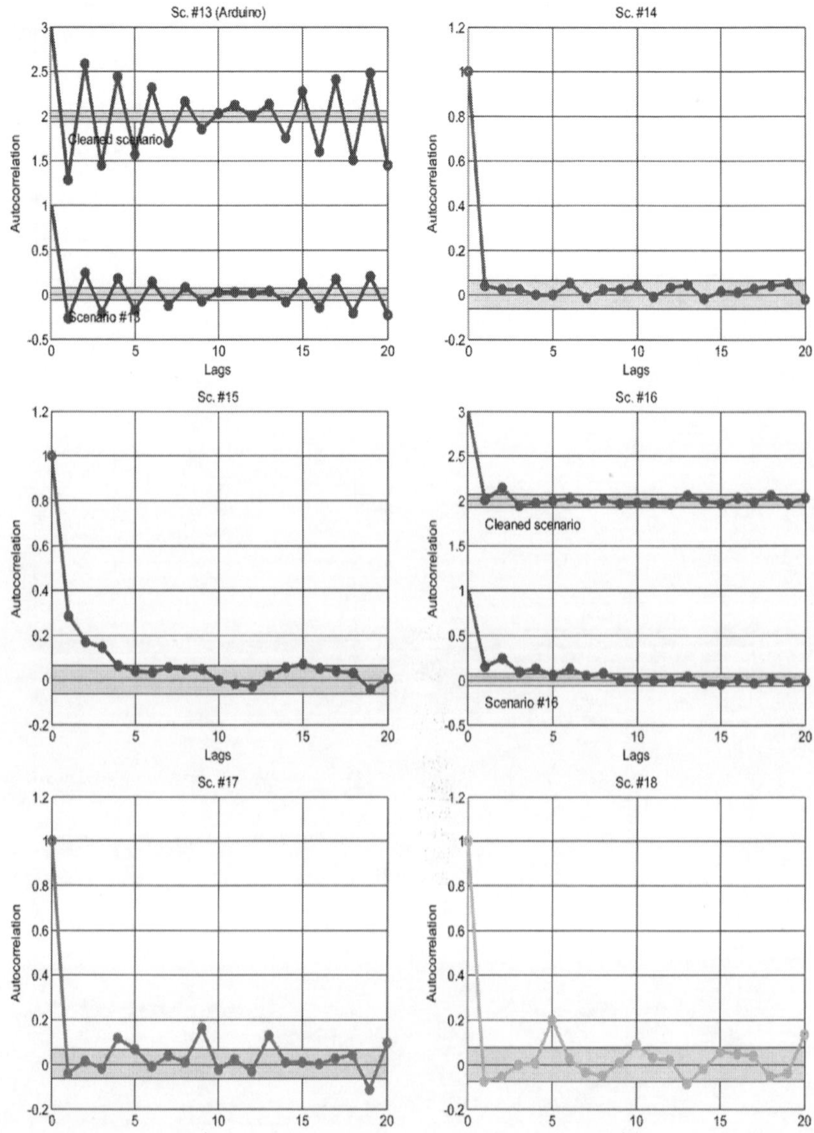

Figure 4. Autocorrelograms of the original scenarios of Figure 2 and of the visually purged scenarios once they are divided into the visual regimes marked there. Observe that in the cases when some lags go above/below the confidence limits, they do not exceed the 5% of the total count of lags (except for scenario #13).

3.1. General Settings

Our main analysis procedure has two steps: (i) a fitting of a certain theoretical marginal model with a large enough set of delays selected from the real scenario; and (ii) a hypothesis test where the rest of the delays of the scenario are used for assessing the suitability of that fitting. We have used even index delay values from each original sample for the fitting step, while odd index values have been selected for the statistical assessment of the goodness of the model (hypothesis testing). This is to preserve as best as possible the same autocorrelations in the temporal sequence of the signal in both steps. Also, using this division we decouple fitting from hypothesis testing, avoiding the reduction in degrees of freedom and allowing us to use tables that exist for some of the statistical tests [36].

Table 3. Mathematical forms of the parametrical pdfs used in our characterization. All of them are considered three-parametrical, being γ the offset, *i.e.*, the lower bound for the delays.

pdf	Constraints
$Erlang(x;\alpha,\beta,\lambda)=\frac{(x-\gamma)^{(\alpha-1)}}{\beta^{\alpha}(\alpha-1)!}e^{\frac{-(x-\gamma)}{\beta}}$	$x,\alpha,\beta,\gamma\in\mathbb{R}\quad x,\alpha,\beta,\gamma>0$
$Gamma(x;\alpha,\beta,\lambda)=\frac{(x-\gamma)^{(\alpha-1)}}{\beta^{\alpha}\Gamma(\alpha)}e^{\frac{-(x-\gamma)}{\beta}}$	$x,\alpha,\beta,\gamma\in\mathbb{R}\quad x,\alpha,\beta,\gamma>0$
$Lognormal(x;\mu,\sigma,\gamma)=\frac{1}{(x-\gamma)\sigma\sqrt{2\pi}}e\frac{-(\ln(x-\gamma)-\mu)^2}{2\sigma^2}$	$\mu,\gamma\in\mathbb{R}\quad x,\sigma^2\geq 0$
$Weibull(x;k,\lambda,\gamma)=\frac{k}{\lambda}\left(\frac{x-\gamma}{\lambda}\right)^{k-1}e^{-((x-\gamma)/\lambda)^k}$	$\lambda,k,\gamma,x\geq 0$
$Exponential(x;\lambda,\gamma)=\lambda e^{-\lambda(x-\gamma)}$	$\lambda,\gamma,x\geq 0$

For modeling the marginal stochastic behaviour of the sequences of delays we have chosen both parametrical (listed inTable 3) and non-parametrical probability density functions (pdfs). We have not included symmetrical distributions like the normal (proposed in some works, such as [14,37]) because of the pronounced skewness of our scenarios. The models we have selected reflect some insights about the characteristics observed in the real data, namely:

- When the system has very short delays—for example when the sensors send few data, as it is the case with the typical range sensors or the odometry of a mobile robot, or when the communications and/or software run really fast—the shape of the statistical distribution of the delays is very close to an exponential

[38]. Actually, exponential distributions are used for modeling inter-arrival times in many situations [39]. In order to make this model more flexible we have also included in our analyses the erlang distribution, which can explain data coming from a sum of k independent exponentially distributed components.

- In the networking literature, long-tailed and heavy-tailed distributions are often assumed for the transmission delays in the network when the delays are not so short. For instance, lognormal distributions have been proposed elsewhere [40]. To our knowledge there is no report in the literature showing that the total delay (*i.e.*, network + client + server) supports this model, but our final results corroborate its suitability for more scenarios than other models; furthermore, when the three additive components of the total delay are split (Section 3.3), or when we use our change detection method (Section 4.2), the lognormal becomes the parametrical model of preference.

- In order to increase the feasibility of the previous mentioned models we have also tested the Weibull distribution, which is a generalization of the exponential. For the same reasons we have included the gamma distribution, which generalizes the erlang and has been proposed for modeling the round trip time of wireless network transmissions [17].

- Non-parametrical models allow for greater flexibility in their shapes than parametrical ones, which has pros and cons: they can fit better more complex behaviours, such as the ones in unprocessed scenarios, but they are more likely to model the variations of a particular sample rather than the underlying stochastic process of the whole signal (*i.e.*, overfitting), being poor models when new samples are gathered. We have included two non-parametrical models in this work: one based on a Gaussian kernel [41], able to model signals with more than one mode (which can be useful for sequences of delays that have more than one regime), and a spline distribution consisting of a cubic polynomial sequence that interpolates a set of control points [42]. The latter is more likely to suffer from overfitting, since it imposes the least rigidity on the data.

- Finally, we have also included in our analyses a hybrid model that draws advantages from both parametrical and non-parametrical representations: a sum or mixture of gaussians (SoG), which is less likely to suffer from overfitting than the spline and is more rigid than the kernel. This model consists of a sum of a given, fixed number of Gaussians (this is different from the Gaussian kernel model, which can have an unlimited number of kernels). We have chosen specifically four Gaussian modes for this SoG, which has been deduced visually from the multimodal behaviours ofFigure 2. Its parameters have to be estimated with a special trust region non-linear optimization algorithm, which is implemented by the Curve Fitting Tool of Matlab [43].

For finding the best fitting for any of these models, a histogram of the data is constructed first with $2n^{2/5}$ cells, where n is the fitting sample size. We use then maximum likelihood estimation (MLE) for each distribution (or the trust algorithm in the case of the SoG). After this fitting, in order to study the goodness-of-fit, we have employed both the Chi-Squared (χ^2) and the Kolmogorov-Smirnov (K-S) significance tests with 0.05 of significance. As it will be shown further on, the former gives worse results in general, particularly due to its higher sensitivity to shorter samples (something that is likely to occur when detecting abrupt changes and separating the sequence of delays into segments, as it is shown in Section 4.2).

3.2. Results for Unprocessed Delays

After running the significance tests, we see in Table 4 that, as expected from the multimodal nature of most scenarios, only a few of them can be explained by parametrical models: #3, #2 and #18 (corresponding to the B class and to a percentage of all delay values in all the scenarios below 12%). The rest of the scenarios do not support the null hypothesis. In the case of non-parametrical models, kernel and spline provide the best results, with above 40% of delay values explained. Non-rejected scenarios are displayed in Figure 5 for illustrative purposes. These results show that, for a parametrical choice, the lognormal or erlang distributions could be preferred when we are not able to detect outliers, bursts and regimes, but these are not good enough and might be applicable to very few scenarios. For a non-parametrical option, the kernel distribution should be the best election. In general non-parametrical models are able to explain scenarios from all classes (A, B, C), given their closest fitting to the particular data in the sample (which leads, on the other hand, to the previously mentioned problem of overfitting). The SoG model obtains worse results than the kernel due to its restriction in the number of modes, while the spline is also worse than the kernel due to its greater overfitting, that produces very good models of the data used for fitting but not in its goodness-of-fit with the delays reserved for hypothesis testing.

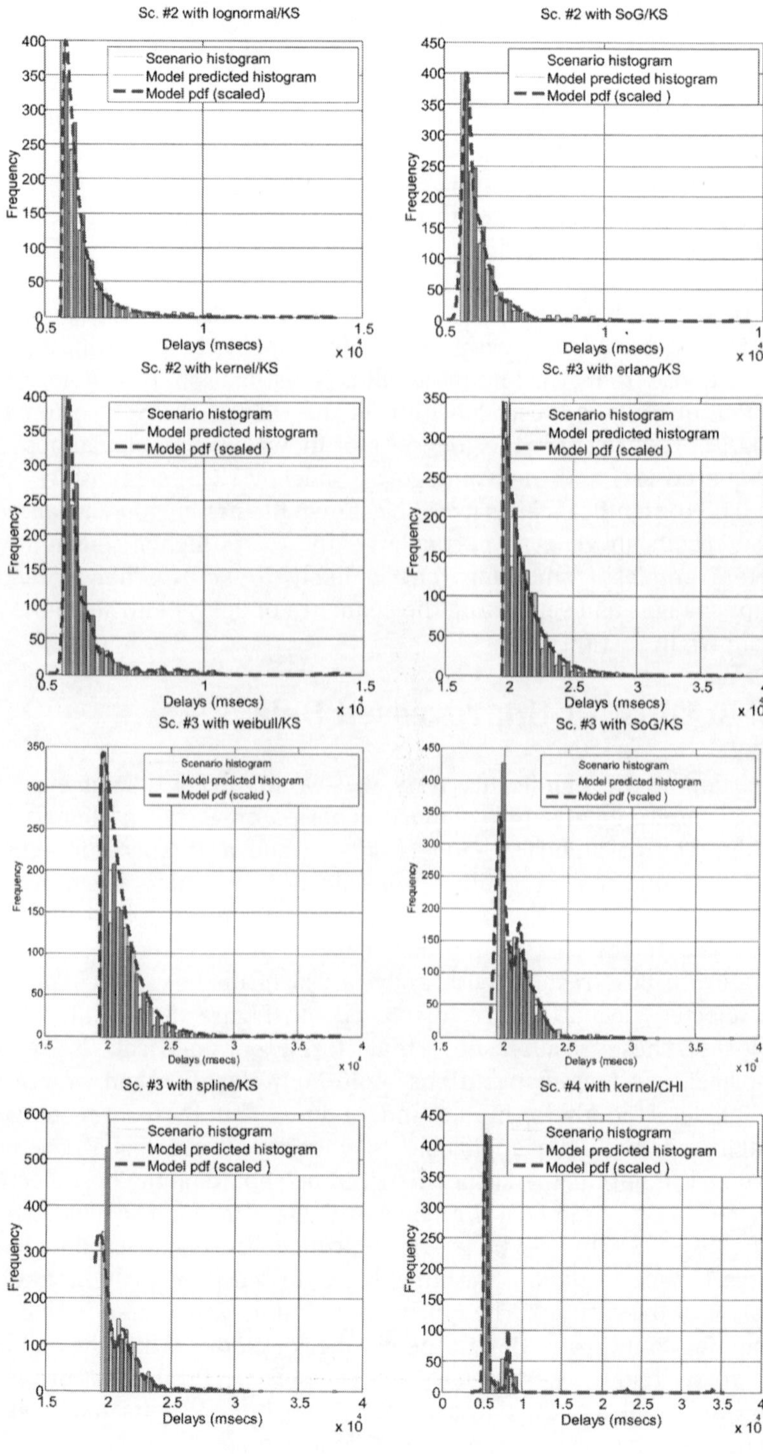
Sc. #2 with lognormal/KS
Scenario histogram
Model predicted histogram
Model pdf (scaled)
Frequency
Delays (msecs)
x 10^4
Sc. #2 with SoG/KS
Scenario histogram
Model predicted histogram
Model pdf (scaled)
Frequency
Delays (msecs)
x 10^4
Sc. #2 with kernel/KS
Scenario histogram
Model predicted histogram
Model pdf (scaled)
Frequency
Delays (msecs)
x 10^4
Sc. #3 with erlang/KS
Scenario histogram
Model predicted histogram
Model pdf (scaled)
Frequency
Delays (msecs)
x 10^4
Sc. #3 with weibull/KS
Scenario histogram
Model predicted histogram
Model pdf (scaled)
Frequency
Delays (msecs)
x 10^4
Sc. #3 with SoG/KS
Scenario histogram
Model predicted histogram
Model pdf (scaled)
Frequency
Delays (msecs)
x 10^4
Sc. #3 with spline/KS
Scenario histogram
Model predicted histogram
Model pdf (scaled)
Frequency
Delays (msecs)
x 10^4
Sc. #4 with kernel/CHI
Scenario histogram
Model predicted histogram
Model pdf (scaled)
Frequency
Delays (msecs)
x 10^4

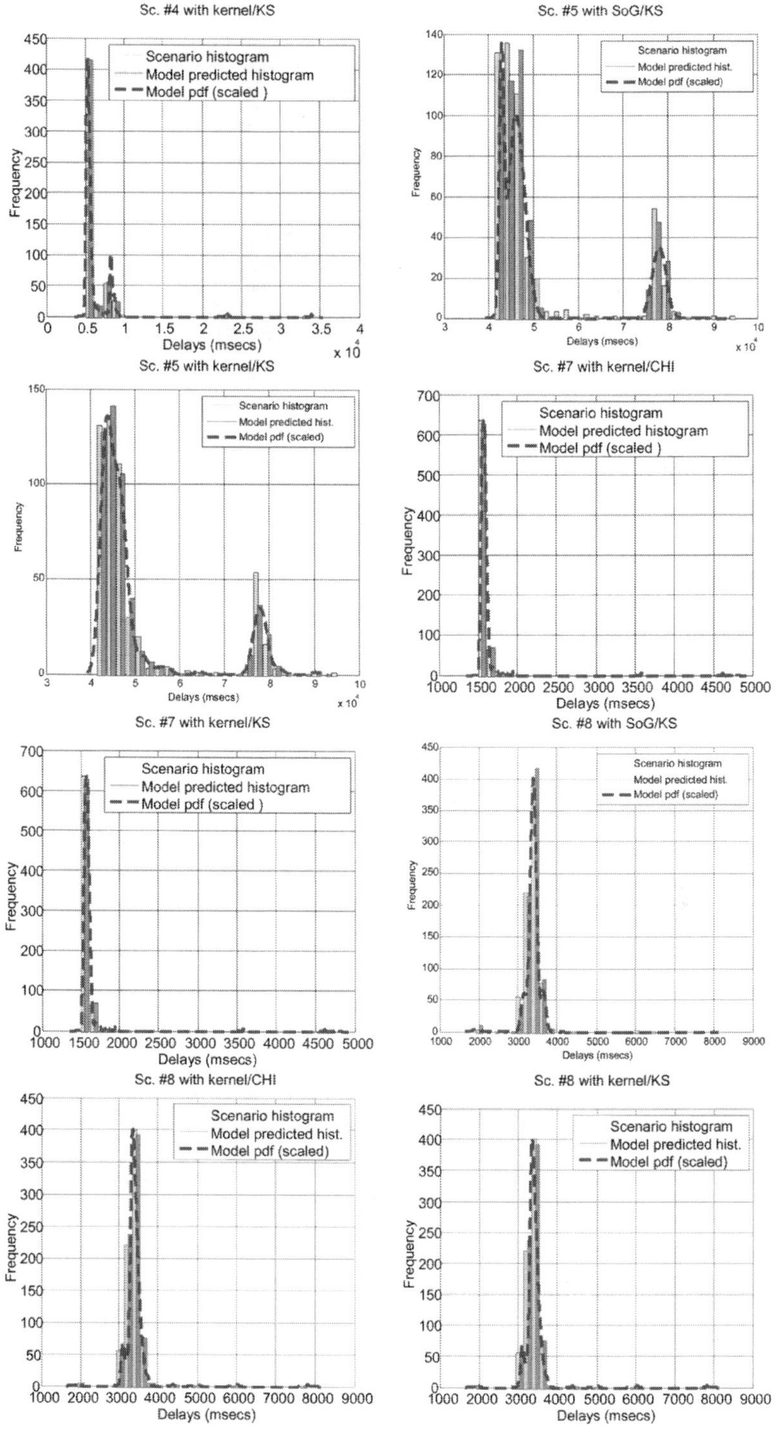

Sc. #4 with kernel/KS
Scenario histogram
Model predicted histogram
Model pdf (scaled)
Frequency
Delays (msecs)
x 10^4
Sc. #5 with SoG/KS
Scenario histogram
Model predicted hist.
Model pdf (scaled)
Frequency
Delays (msecs)
x 10^4
Sc. #5 with kernel/KS
Scenario histogram
Model predicted hist.
Model pdf (scaled)
Frequency
Delays (msecs)
x 10^4
Sc. #7 with kernel/CHI
Scenario histogram
Model predicted histogram
Model pdf (scaled)
Frequency
Delays (msecs)
Sc. #7 with kernel/KS
Scenario histogram
Model predicted histogram
Model pdf (scaled)
Frequency
Delays (msecs)
Sc. #8 with SoG/KS
Scenario histogram
Model predicted hist.
Model pdf (scaled)
Frequency
Delays (msecs)
Sc. #8 with kernel/CHI
Scenario histogram
Model predicted hist.
Model pdf (scaled)
Frequency
Delays (msecs)
Sc. #8 with kernel/KS
Scenario histogram
Model predicted hist.
Model pdf (scaled)
Frequency
Delays (msecs)

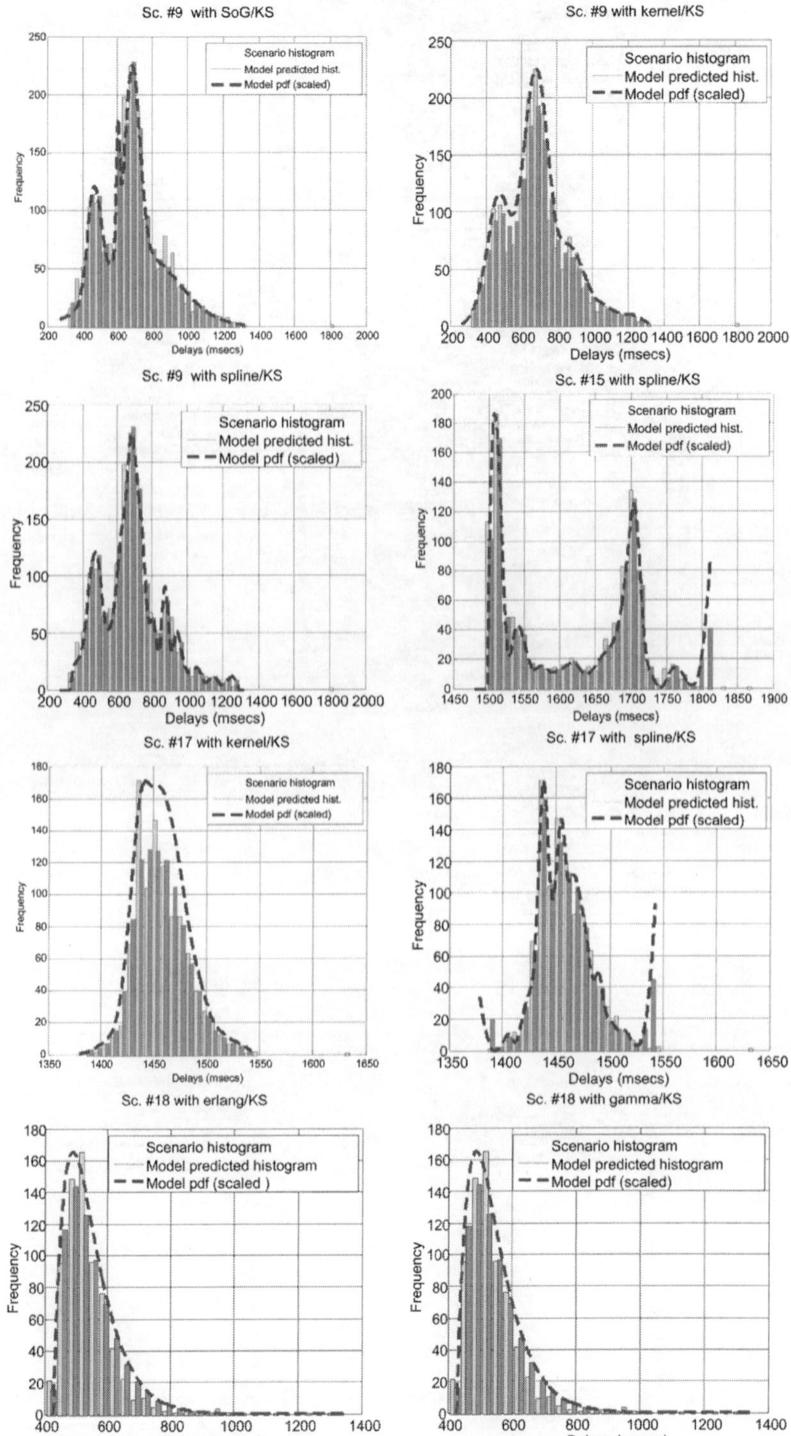
Sc. #9 with SoG/KS
Sc. #9 with kernel/KS
Sc. #9 with spline/KS
Sc. #15 with spline/KS
Sc. #17 with kernel/KS
Sc. #17 with spline/KS
Sc. #18 with erlang/KS
Sc. #18 with gamma/KS
Scenario histogram
Model predicted hist.
Model pdf (scaled)
Model predicted histogram
Frequency
Delays (msecs)

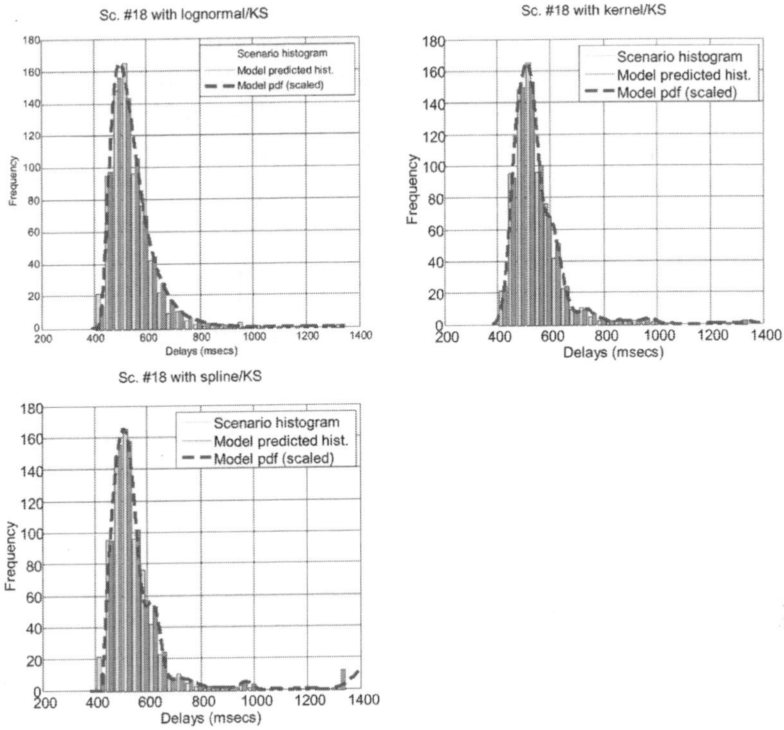

Figure 5. Illustrative histograms and fittings of the models that do not reject one or both hypothesis tests according to Table 4.

In summary, the higher percentage of non-rejected samples of all scenarios (bold numbers in Table 4) corresponds to a non-parametrical distribution, the kernel model, although with less than 50% of the total delays that have been tested. This can be further improved without filtering non-linearities in the sequence of delays, as it will be explained in the next section.

3.3. Results for Unprocessed Split Delays

To obtain better adjustments of the marginal models without detecting and processing bursts and regimes yet, we can divide the complete timing behaviour of the sensory loop into three different terms that are added to obtain the total delay: Server time, client time and network time. They can be modeled separately and perhaps more accurately.

Table 4. Non-rejected hypotheses for each unprocessed scenario, indicating which class it belongs to and the percentage of non-rejected delay values (in bold), that is, the proportion of delays explained by that model with respect to all delay values of all scenarios tested. This percentage is calculated independently of the length of each scenario for a fairer comparison.

	Exponential (*p*-Value)	Erlang (*p*-Value)	Gamma (*p*-Value)	Lognormal (*p*-Value)	Weibull (*p*-Value)	SoG (*p*-Value)	Kernel (*p*-Value)	Spline (*p*-Value)
χ^2	-	-	-	-	-	-	#4B2(0.14) #7B2(0.51) #8C(0.16) **16.7%**	-
KS	-	#3A/B2(0.03) #18B1(0.06) **11.1%**	#18B1(0.08) **5.6%**	#2B1(0.07) #18B1(0.03) **11.1%**	#3A/B2(0.03) **5.6%**	#2B1(0.13) #3A/B2(0.05) #5B2(0.04) #8C(0.39) #9A(0.12) **27.8%**	#2B1(0.30) #4B2(0.44) #5B2(0.21) #7B2(0.05) #8C(0.50) #9A(0.04) #17A(0.07) #18B1(0.40) **44.4%**	#3A/B2(0.28) #9A(0.25) #15A(0.03) #17A(0.18) #18B1(0.52) **27.8%**

Table 5. Condensed comparative table of non-rejected scenarios for both unprocessed (U) and split (S) scenarios, with parametrical and non-parametrical distributions (the unprocessed part is the same as Table 4).

	Exponential (*p*-Value)	Erlang (*p*-Value)	Gamma (*p*-Value)	Lognormal (*p*-Value)	Weibull (*p*-Value)	SoG (*p*-Value)	Kernel (*p*-Value)	Spline (*p*-Value)
χ^2 (U)	-	-	-	-	-	-	#4B2(0.14) #7B2(0.51) #8C(0.16) **16.7%**	-
KS (U)	-	#3A/B2(0.03) #18B1(0.06) **11.1%**	#18B1(0.08) **5.6%**	#2B1(0.07) #18B1(0.03) **11.1%**	#3A/B2(0.03) **5.6%**	#2B1(0.13) #3A/B2(0.05) #5B2(0.04) #8C(0.39) #9A(0.12) **27.8%**	#2B1(0.30) #4B2(0.44) #5B2(0.21) #7B2(0.05) #8C(0.50) #9A(0.04) #17A(0.07) #18B1(0.40) **44.4%**	#3A/B2(0.28) #9A(0.25) #15A(0.03) #17A(0.18) #18B1(0.52) **27.8%**
χ^2 (S)	-	-	-	-	-	-	#4srvB2(0.87) #4netB2(0.47) #4cliB2(0.06) #6netB2(0.33) #7srvB2(0.08) #18cliB1(0.07) **28.6%**	-

	Exponential (*p*-Value)	Erlang (*p*-Value)	Gamma (*p*-Value)	Lognormal (*p*-Value)	Weibull (*p*-Value)	SoG (*p*-Value)	Kernel (*p*-Value)	Spline (*p*-Value)
KS (S)	-	-	#18netB1(0.05) **4.8%**	#4netB2(0.04) #5cliB2(0.03) #18cliB1(0.06) #18netB1(0.07) **19%**	-	#4netB2(0.17) #5srvB2(0.17) #5netB2(0.31) #18cliB1(0.05) #18netB1(0.22) **23.8%**	#4srvB2(0.23) #4netB2(0.37) #5srvB2(0.24) #5netB2(0.12) #5cliB2(0.80) #6srvB2(0.07) #6netB2(0.60) #7srvB2(0.42) #8srvC(0.48) #8netC(0.17) #8cliC(0.17) #18cliB1(0.03) #18netB1(0.50) #18srvB1(0.24) **66.7%**	#4cliB2(0.13) #5srvB2(0.58) #5cliB2(0.05) #18cliB1(0.04) #18netB1(0.59) #18srvB1(0.73) **28.6%**

These split times can be measured without much intervention in many networked systems, just by time-stamping at the client interface and at the server side the sensory data being transmitted, which neither involves a significant overload nor requires the use of distributed clock synchronization algorithms. However, this splitting is not always possible. There are a number of reasons for this: (i) the server side might not allow us to time-stamp messages because it is a closed implementation; (ii) if the format of the messages carrying the requested sensor data cannot be modified, they cannot include the time-stamps; and (iii) the clock precision on the server side might not be good enough to distinguish short server delays, which are really common—we cannot allow for zero delays in our approach because of the impossibility of many distributions to explain their occurrence: lognormal, weibull, erlang and gamma are limited to model strictly positive delays in most or all of their parameter configurations.

Due to the previous situations, we have been able to collect split information only from scenarios #4–#8, #13 and #18, summarising the results in Table 5. In general, parametric distributions do not fit better when we split the delays, except for the lognormal distribution. However, when non-parametrical distributions are set up as null hypotheses, the K-S test does not reject the null hypothesis of kernel for 66.7% of the total of delays, which is a really good result. Thus, in situations when we are not able to detect non-linearities, the kernel model should be the preferred approximation, if augmented with this simple splitting of delays.

In Table 5 and Figure 6 it can be seen how the different parts of the delays can be more or less predominant in the total time, and most of them can be modeled with our marginal approach. This serves to demonstrate that in a networked telerobot, taking into account only the delay produced by the network transmission of data (as many approaches in literature do) is not appropriate.

4. PROCESSING DELAYS FOR THE DETECTION OF REGIMES AND BURSTS

We have shown in Section 3 the great influence of the non-linearities of the delay signal, especially abrupt changes, in the possibility of modeling these delays with both unimodal and multimodal marginal pdfs, even when they are split into their most relevant additive components. We explore now the effects of detecting and ruling out such abrupt characteristics. In Section 4.1 we explain a method specifically designed for this work due to its accurateness/complexity trade-off, able to detect abrupt changes in the signal and therefore separate the temporal sequence into parts that do not contain those non-linearities. This method cannot guarantee, however, that the resulting segments pass the hypothesis test, only that they should be separately modeled. Section 4.2 presents its results when applied to the delays of our scenarios. Section 4.3 complements the change detection analysis through the use of a previously presented algorithm for segmentation of the signal [23] that, on the one hand, is much more involved computationally (thus, without the inclusion of especial techniques for improving computational efficiency it does not satisfies our requirement for low computational cost), but, on the other hand, tries to find only segments of the signal that can be explained by the model at hand. The results of this method are gathered in a table that serves as a near-optimal segmentation of the signal for the goal of modeling it with each probability distribution. Finally, Section 4.4 completes our analyses through the use of another two statistic tools, Q-Q graphs and probability difference graphs, which serve to draw additional conclusions on our modeling problem.

4.1. Efficient Abrupt Change Detection

Our approach to detect abrupt changes in the signal efficiently is based on identifying outliers. Although the precise definition of outlier is quite subjective, in our context it will be any value that is not likely to be produced by the model we have constructed up to that time: in that way we will use sequences of consecutive outliers to define abrupt changes in the underlying distribution, and thus, to separate near-stationary sequences of delays.

The literature in outlier detection is extensive (see for example [44,45]). In principle we are interested in detecting them when using both parametrical and non-parametrical models. Some outlier detection approaches could be used for that because they do not impose any particular probabilistic model for the data, for instance those that are based on clustering techniques [46,47], *i.e.*, on considering outliers those values that form small clusters separated from the main data; other approaches search for values that may be consistent with the global set of data but are not with their local neighborhoods, called spatial outliers [48]. Unfortunately, these distribution-independent techniques are usually computationally involved. Some of them are based on local distance measures, and usually designed to manage high-dimensional spaces efficiently in data-mining [49–51]. This presents another problem: they require large sets of data—long sequences of delays—which is not typically possible in our framework and the kind of scenarios found in this work. In summary, due to the lack of existing methods that are both efficient and able to deal with small samples in a non-parametrical setting, we have focused our analysis firstly on providing a solution for parametrical models only, and secondly, based on the obtained results, on deciding whether further research on non-parametrical approaches is worth. As explained later on, we report here good results with a basic method that works only with parametrical models, thus we have left for future work the study of non-parametrical ones.

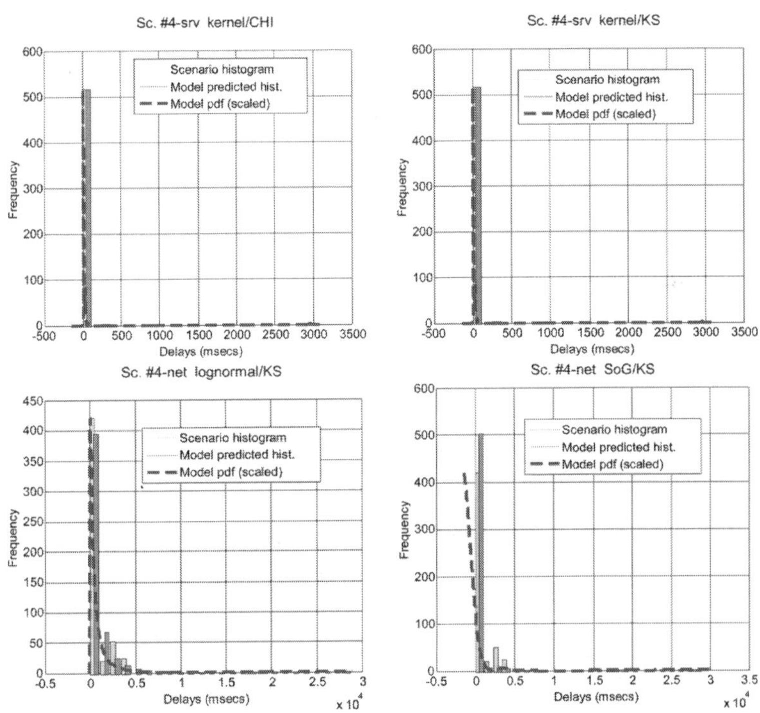

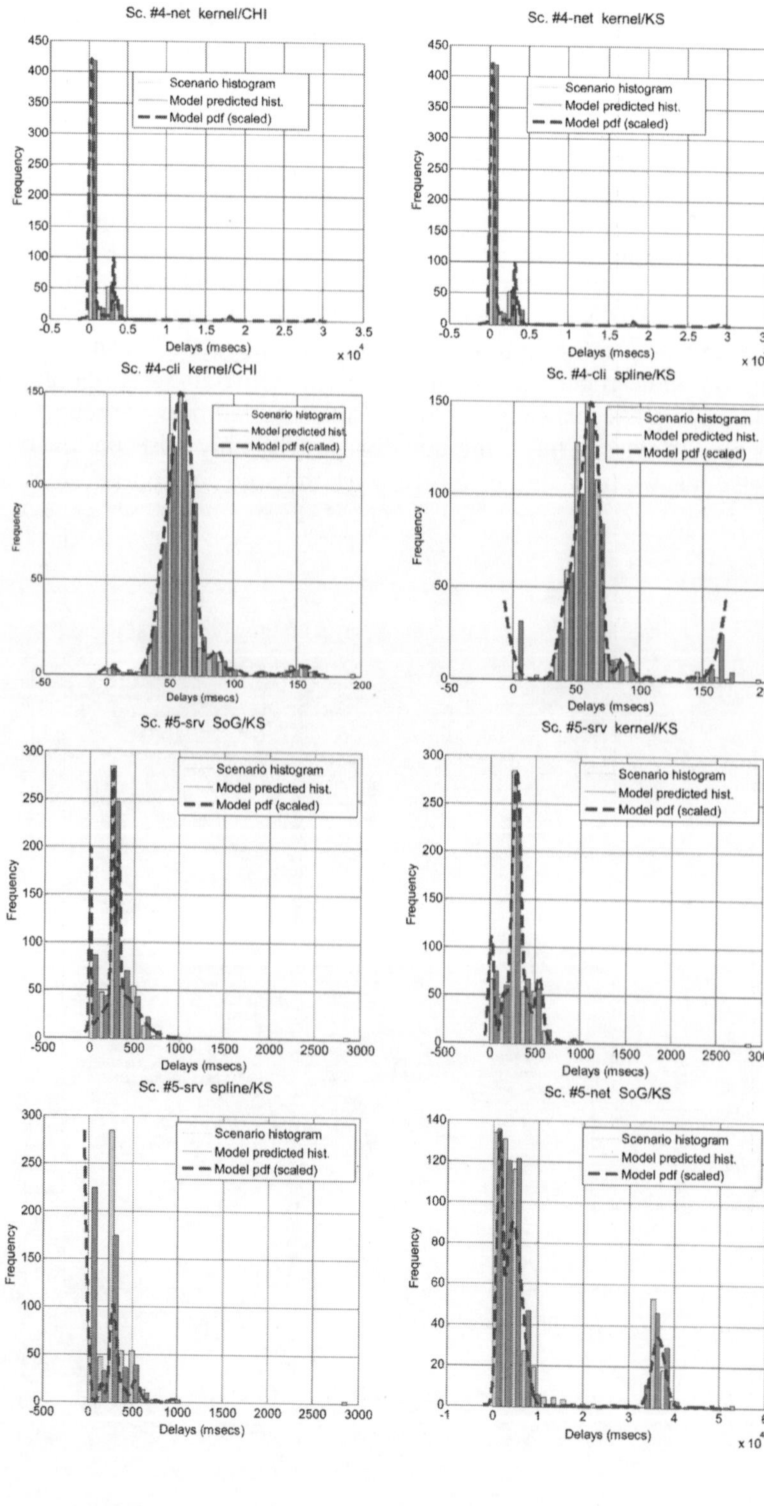

Sc. #4-net kernel/CHI
Scenario histogram
Model predicted hist.
Model pdf (scaled)
Frequency
Delays (msecs)
x 10^4
Sc. #4-net kernel/KS
Scenario histogram
Model predicted hist.
Model pdf (scaled)
Frequency
Delays (msecs)
x 10^4
Sc. #4-cli kernel/CHI
Scenario histogram
Model predicted hist
Model pdf s(called)
Frequency
Delays (msecs)
Sc. #4-cli spline/KS
Scenario histogram
Model predicted hist.
Model pdf (scaled)
Frequency
Delays (msecs)
Sc. #5-srv SoG/KS
Scenario histogram
Model predicted hist.
Model pdf (scaled)
Frequency
Delays (msecs)
Sc. #5-srv kernel/KS
Scenario histogram
Model predicted hist.
Model pdf (scaled)
Frequency
Delays (msecs)
Sc. #5-srv spline/KS
Scenario histogram
Model predicted hist.
Model pdf (scaled)
Frequency
Delays (msecs)
Sc. #5-net SoG/KS
Scenario histogram
Model predicted hist.
Model pdf (scaled)
Frequency
Delays (msecs)
x 10^4

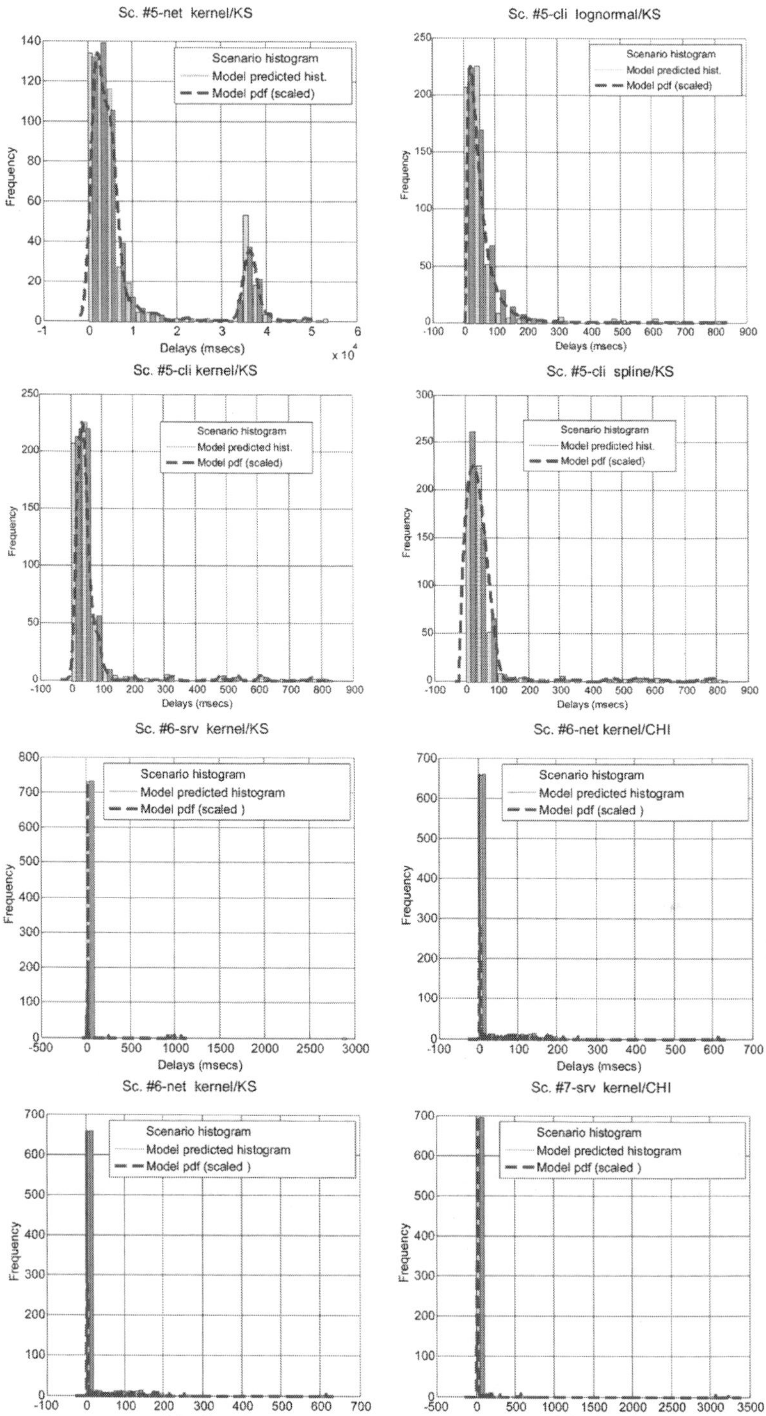
Sc. #5-net kernel/KS
Scenario histogram
Model predicted hist.
Model pdf (scaled)
Frequency
Delays (msecs)
x 10^4
Sc. #5-cli lognormal/KS
Scenario histogram
Model predicted hist.
Model pdf (scaled)
Frequency
Delays (msecs)
Sc. #5-cli kernel/KS
Scenario histogram
Model predicted hist.
Model pdf (scaled)
Frequency
Delays (msecs)
Sc. #5-cli spline/KS
Scenario histogram
Model predicted hist.
Model pdf (scaled)
Frequency
Delays (msecs)
Sc. #6-srv kernel/KS
Scenario histogram
Model predicted histogram
Model pdf (scaled)
Frequency
Delays (msecs)
Sc. #6-net kernel/CHI
Scenario histogram
Model predicted histogram
Model pdf (scaled)
Frequency
Delays (msecs)
Sc. #6-net kernel/KS
Scenario histogram
Model predicted histogram
Model pdf (scaled)
Frequency
Delays (msecs)
Sc. #7-srv kernel/CHI
Scenario histogram
Model predicted histogram
Model pdf (scaled)
Frequency
Delays (msecs)

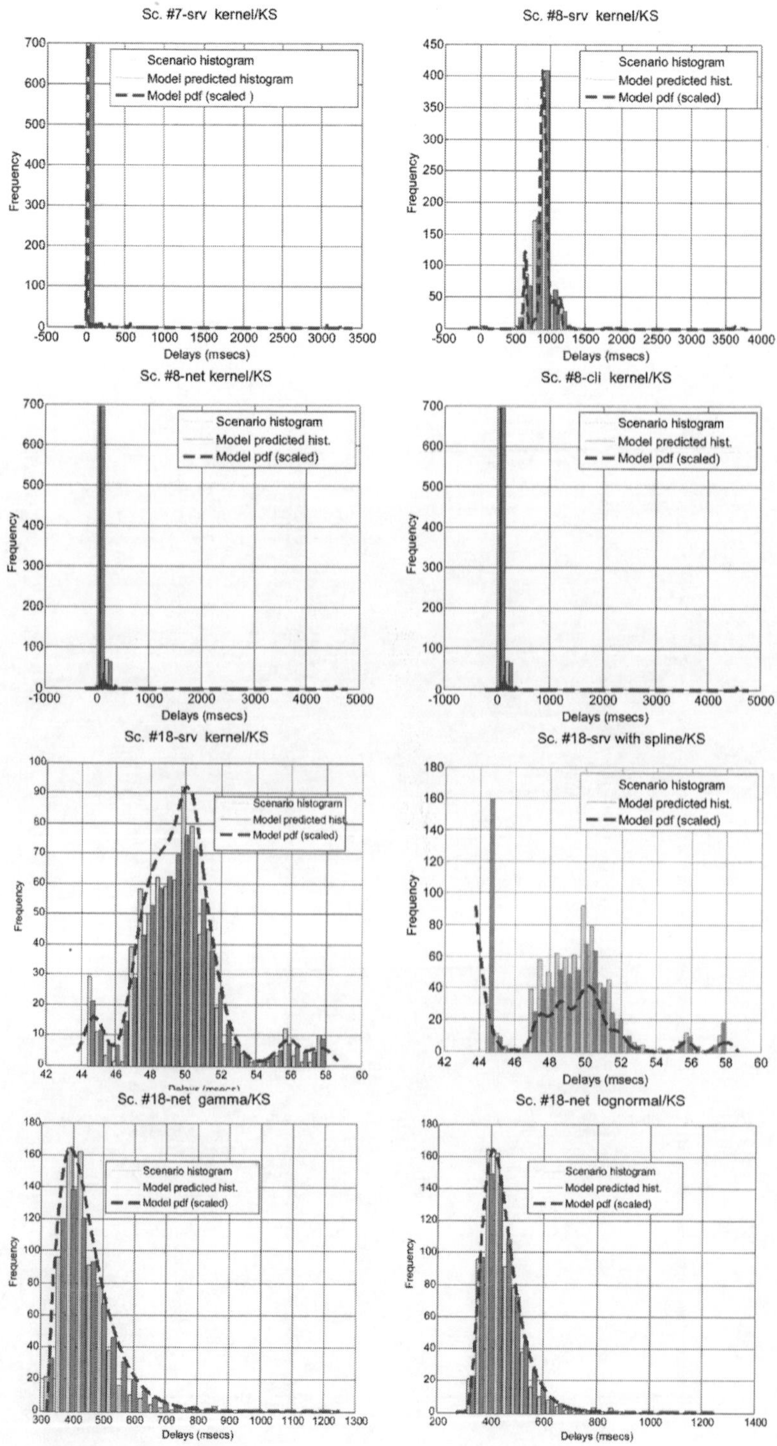
Sc. #7-srv kernel/KS
Scenario histogram
Model predicted histogram
Model pdf (scaled)
Frequency
Delays (msecs)
Sc. #8-srv kernel/KS
Scenario histogram
Model predicted hist.
Model pdf (scaled)
Frequency
Delays (msecs)
Sc. #8-net kernel/KS
Scenario histogram
Model predicted hist.
Model pdf (scaled)
Frequency
Delays (msecs)
Sc. #8-cli kernel/KS
Scenario histogram
Model predicted hist.
Model pdf (scaled)
Frequency
Delays (msecs)
Sc. #18-srv kernel/KS
Scenario histogram
Model predicted hist
Model pdf (scaled)
Frequency
Sc. #18-srv with spline/KS
Scenario histogram
Model predicted hist.
Model pdf (scaled)
Frequency
Delays (msecs)
Sc. #18-net gamma/KS
Scenario histogram
Model predicted hist.
Model pdf (scaled)
Frequency
Delays (msecs)
Sc. #18-net lognormal/KS
Scenario histogram
Model predicted hist.
Model pdf (scaled)
Frequency
Delays (msecs)

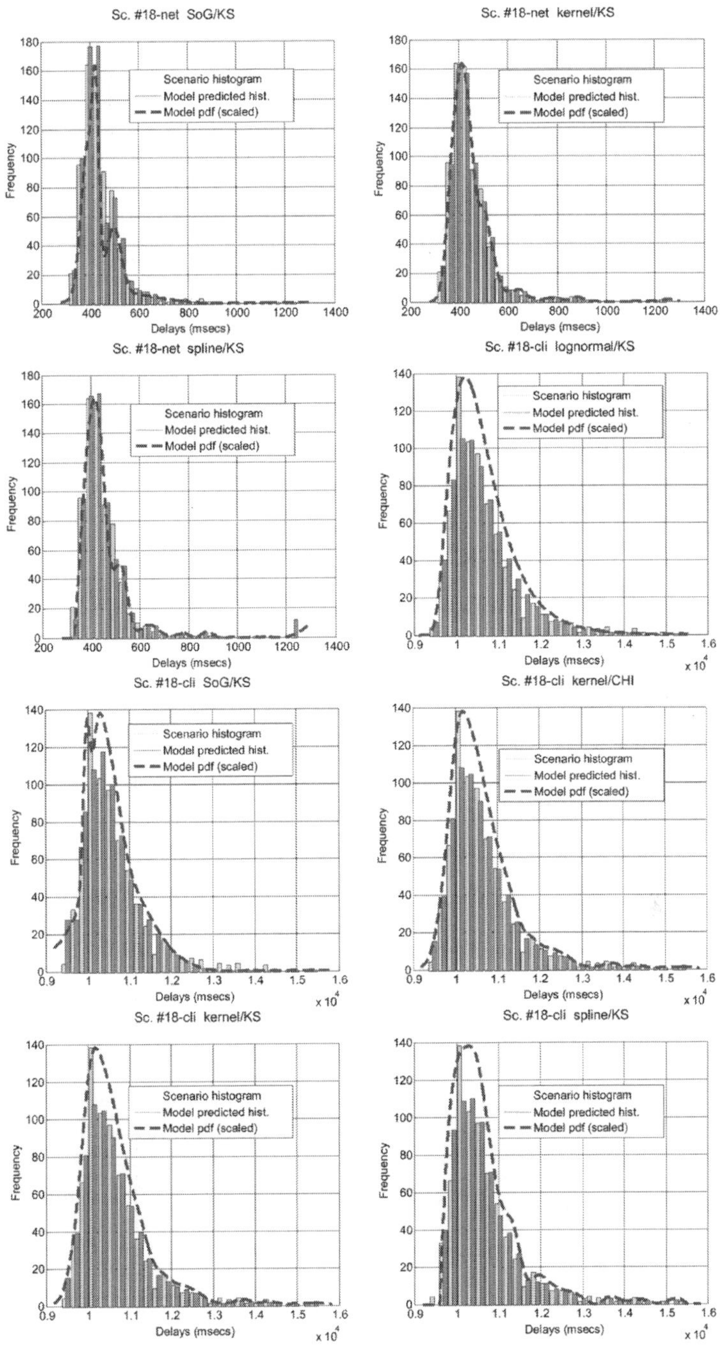

Figure 6. Illustrative histograms and fittings for the scenarios that can be modeled successfully after the splitting of delays (see Table 5).

The method we propose for detecting abrupt changes with low computational cost is based on the basic probability theory idea of calculating the probability that a value has not been drawn from a known model, or, in our case, that it lies in an "outlier-region" defined to be in the right-tail of the current model (by a given significance level α). The adaptation of this basic confidence interval approach to our parametrical models has provided an efficient change detection procedure that has good results even with short sequences of delays. It is described in the following.

We gather first a sequence of 40 delay values (called *blind period*) to make an initial, minimally suitable model of the data, using 20 delays for the fitting and 20 for the test. From that point on, each new delay value is considered as an outlier or not depending on the following. A fitting of the collected delays with the chosen parametrical distribution is found out, as before, with the MLE. This estimation also provides 95% confidence intervals for the parameters of the model (this is the reason why our method only works with parametrical distributions). The range of delay values that determines if the new delay is classified as an outlier or not is delimited then by the estimated offset of the distribution on the left and a bound $x_\alpha$on the right. This x_α, in turn, is calculated by finding the x-axis value of the chosen distribution that satisfies the predefined significance level:

$$1 - cdf(x = x_a) = \alpha \qquad (1)$$

We solve the above equation for x_α, thus x_α becomes a function of α and of the estimated parameters of the distribution, which have several possible values if we use the 95% confidence interval obtained in their estimation. In order to get a value of x_α that produces as much outlier detection as possible (*i.e.*, minimizes x_α), we have to select values for that available set of distribution parameters. The ones that accomplish this can be picked up by studying analytically the minimum of the equation for x_α, as it is summarised in Table 6: there we have shown bi-parametrical distributions, because the fitting process estimates firstly the offset (without 95% confidence interval) and, through a bi-parametrical fitting tool of Matlab, the rest of the parameters and their 95% confidence intervals. Exceptionally, for the case of the lognormal distribution, we also have the asymptotic variance of the offset [52], and thus, a 95% confidence interval for that parameter too.

Table 6. Confidence interval parameter selection that minimizes x_α depending on each parametrical distribution assumption.

Distribution (bi-parametrical; the offset has no 95% c.i.)	Parameters provided by estimation, with 95% c.i.	x_α	Confidence interval parameter selection * that minimizes x_α
$Erlang(x;\alpha,\beta)$	a_L, a_U, b_L, b_U	$b\gamma_{inc}^{-1}((1-\alpha)(a-1)!/\Gamma(a),a)$	a_L, b_L
$Gamma(x;\alpha,\beta)$	a_L, a_U, b_L, b_U	$\mu\gamma_{inc}^{-1}((1-\alpha)(k-1)!/\Gamma(k),k)$	a_L, b_L
$Lognormal(x;\mu,\sigma)$	μ_L, μ_U, σ_L, σ_U	$e^{\mu-\beta\sigma\sqrt{2}}$	$if \begin{cases} \beta \le 0, \mu_L, \sigma_L \\ \beta > 0, \mu_L, \sigma_L \end{cases}$
$Weibull(x;k,\lambda)$	k_L, k_U, λ_L, λ_U	$\lambda(-\ln\alpha))^{1/k}$	λ_L, k_U
$Exponential(x;\lambda)$	λ_L, λ_U	$-\lambda\ln(\alpha)$	λ_L

* $\beta = erfcinv(2(1-\alpha))$, where `erfcinv` is the inverse of the complementary error function. γ_{inc}^{-1} is the gamma incomplete inverse function.

If the new delay value lies outside the calculated range, it should be treated as an outlier. If this situation persists for five to forty consecutive outliers, all of them are discarded from the scenario as a burst. Forty consecutive outliers indicate a change of regime. If that occurs, these forty delays will be included in the beginning of the new regime, *i.e.*, in a new blind period. Finally, isolated—singleton—outliers will be treated as part of the current distribution, but if the consecutive number of outliers is higher than one and less than five, these outliers will be discarded.

4.2. Results for Processed Delays

In Table 7 we show how the confidence interval method explained before produces better fits than the splitting of delays (recall Table 5) in all parametrical models. This method, with these parametrical distributions, explains well classes A and B (it does not reject the null hypothesis of the tests), especially when we use the lognormal, including the mostly deterministic scenario #13 corresponding to the Arduino system. As it already happened with split data, erlang suffers both from split and change detection, which, along with their intrinsic reduction in the size of the sample, are procedures that call for more flexible models (remember that one of the parameters of the erlang is forced to be a natural number; the gamma, which is a generalization of the erlang, does not have that limitation). Notice that our previous categorisation {A, B, C} is reflected in these final results: scenarios of class B (they have no trend) are all explained by the methods, even with models that have low rate of success with the data of other classes, while classes A and C are more problematic (but the lognormal is able to model many scenarios from class A, whose trends are successfully reduced by the confidence interval method). It is observed again that the χ^2 test provides less model adjustments than the K-S due to its sensitivity to the reduction of sample size produced by both splitting and change detection methods.

Figure 7 illustrates the results for the scenarios that have been modeled successfully. On the top plot of each figure we have drawn the models found for the different segments (regimes) that have been separated by the confidence interval method. Bursts and outliers are marked with black asterisks.

Table 7. Comparative table of non-rejected scenarios for unprocessed (U), split (S) and confidence interval method (C) with parametrical distributions (unprocessed and split values are copied from tables 4 and 5 respectively).

	Exponential (p-Value)	Erlang (p-Value)	Gamma (p-Value)	Lognormal (p-Value)	Weibull (p-Value)
χ^2 (U)	-	-	-	-	-
KS (U)	-	#3A/B2(0.03) #18B1(0.06) **11.1%**	#18B1(0.08) **5.6%**	#2B1(0.07) #18B1(0.03) **11.1%**	#3A/B2(0.03) **5.6%**
χ^2 (S)	-	-	-	-	-
KS (S)	-	-	#18netB1(0.05) **4.8%**	#4netB2(0.04) #5cliB2(0.03) #18cliB1(0.06) #18netB1(0.07) **19%**	-
χ^2 (C)	-	-	-	#17A(0.17) **0.7%**	-
	Exponential (p-value)	**Erlang (p-value)**	**Gamma (p-value)**	**Lognormal (p-value)**	**Weibull (p-value)**
KS (C)	#4B2(0.19) #5B2(0.47) **1.4%**	#5B2(2;0.09) **5.1%**	#4B2(0.45) #5B2(2;0.32) #18B1(0.15) **10.7%**	#1A(14:0.27) #3A/B2(2;0.35) #4B2(2;0.30) #5B2(2;0.44) #7B2(4;0.28) #8C(0.57) #9A(3;0.30) #13C(4;0.08) #16B2(0.10) #17A(14;0.44) #18B1(9;0.50) **23.5%**	#5B2(0.42) #17A(0.17) **6%**

4.3. Near-Optimal Abrupt Change Detection

In this section we include the results obtained when we use a different change detection algorithm, reported elsewhere as the *stateless* algorithm [23], to separate the delay sequence into almost stationary segments. The algorithm, that we will call here the near-optimal method, relies in executing the complete hypothesis testing each time a new delay value is gathered from the system, which makes its computational burden much higher than the one of the method introduced in Section 4.1. Therefore, this one is included here more as a limit of the results that any detection algorithm can provide with our marginal distribution approach than as an alternative to be implemented in very low computational power system.

The algorithm is quite intuitive. Essentially, it firstly makes up a regime of a minimum size (we use 40 in this work to be in the same conditions as the confidence interval method); when such a minimal regime is collected, it uses the hypothesis test: a rejection moves that regime one delay value on—*i.e.*, we request new sensory data and obtain a new delay measurement—forgetting the oldest delay value of the current regime, while a non-rejection makes the algorithm to try to enlarge the regime with the next delay values, until one is added that makes the test to reject it; when such an enlarged regime is finally rejected, its longest non-rejected portion is kept as a definitive regime of the scenario. It is clear that with this algorithm it is very likely to obtain segments of the signal that can be explained statistically by the model at hand, since they are constructed specifically for that purpose. The algorithm does not find such segments when the scenario is not compatible with the model.

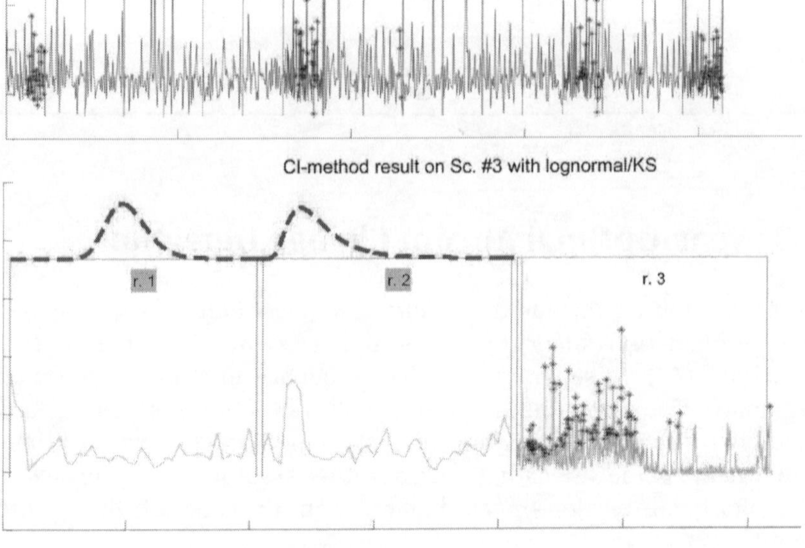

Cl-method result on Sc. #4 with exponential/KS

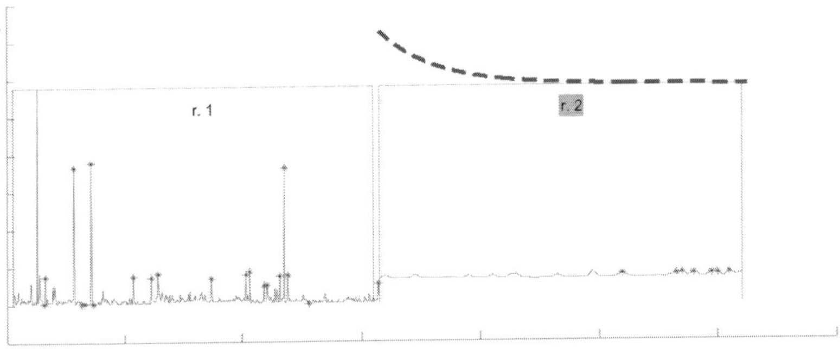

Cl-method result on Sc. #4 with gamma/KS

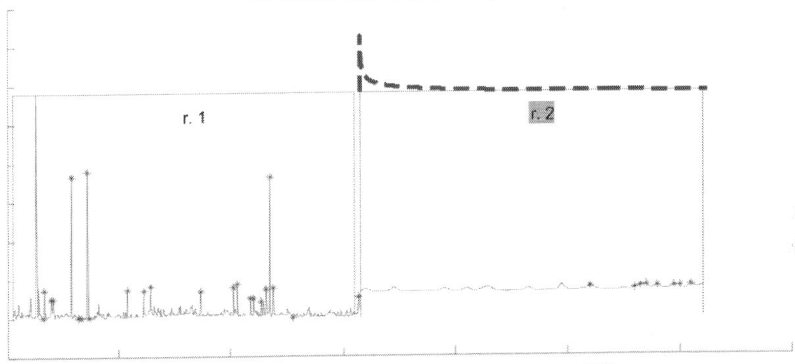

Cl-method result on Sc. #4 with lognormal/KS

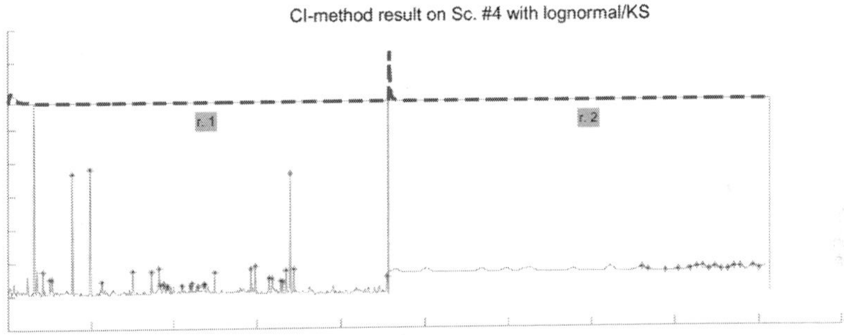

Cl-method result on Sc. #5 with exponential/KS

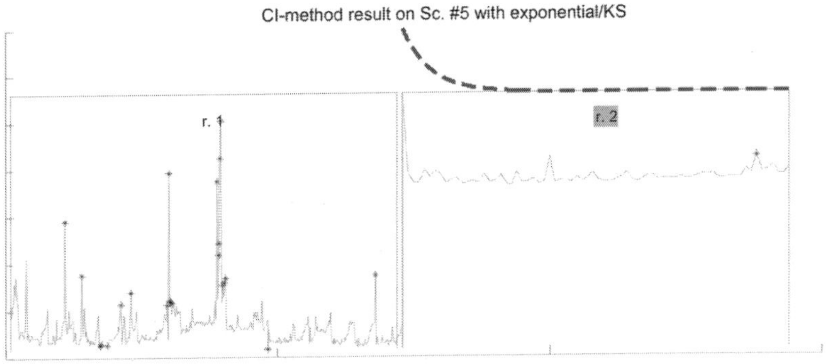

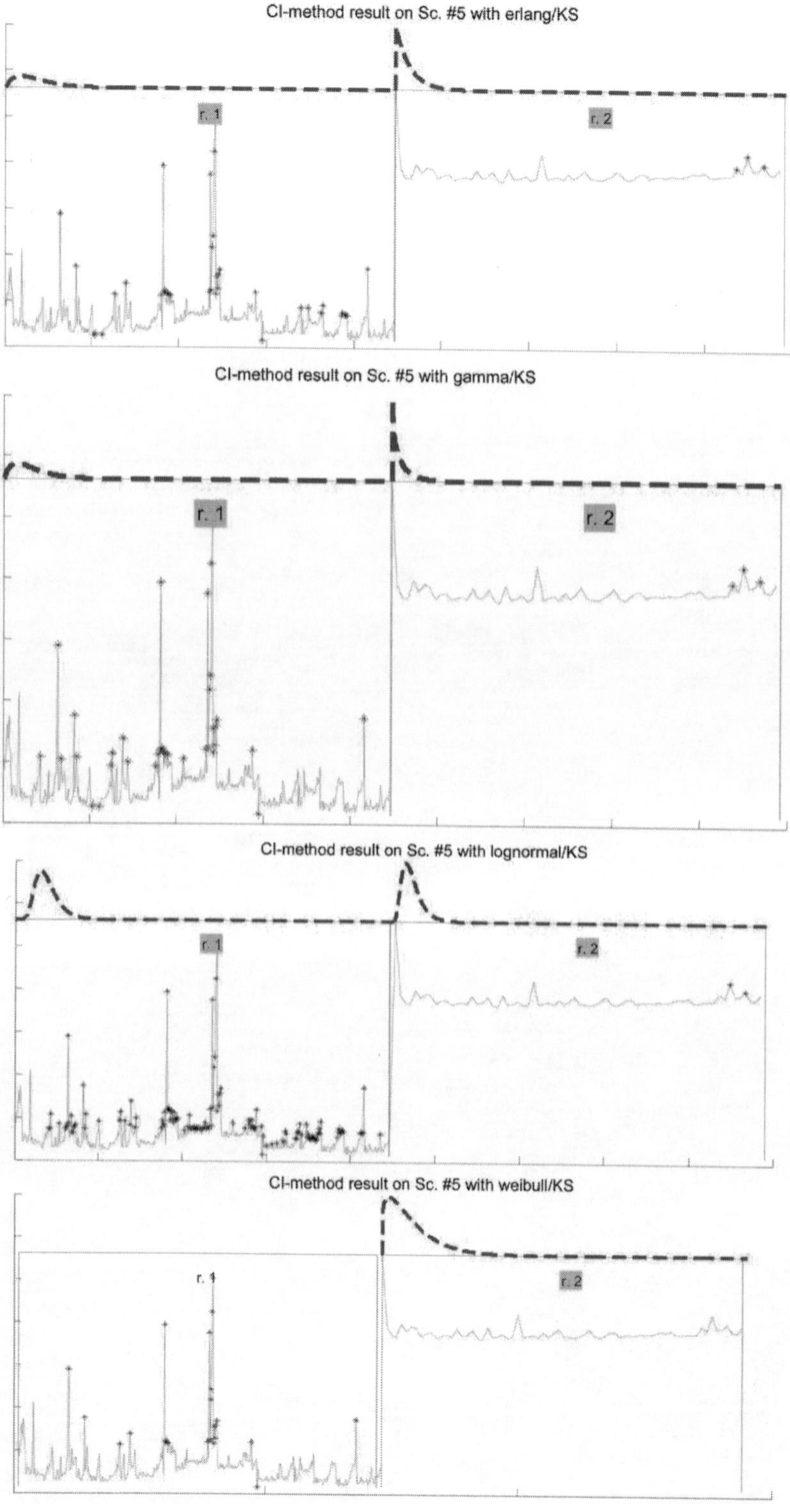
CI-method result on Sc. #5 with erlang/KS
r. 1
r. 2
CI-method result on Sc. #5 with gamma/KS
r. 1
r. 2
CI-method result on Sc. #5 with lognormal/KS
r. 1
r. 2
CI-method result on Sc. #5 with weibull/KS
r. 1
r. 2

Cl-method result on Sc. #7 with lognormal/KS

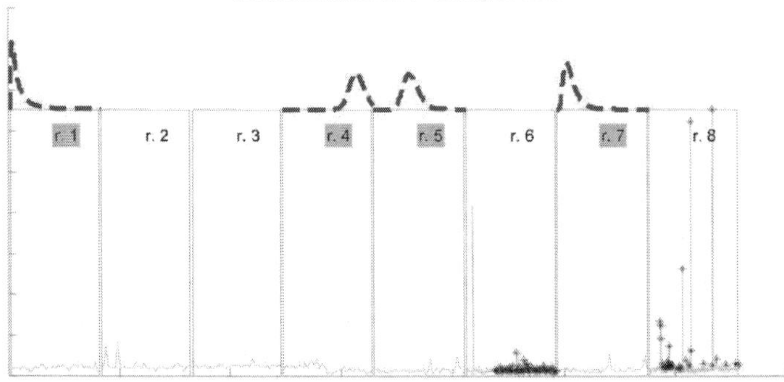

Cl-method result on Sc. #8 with lognormal/KS

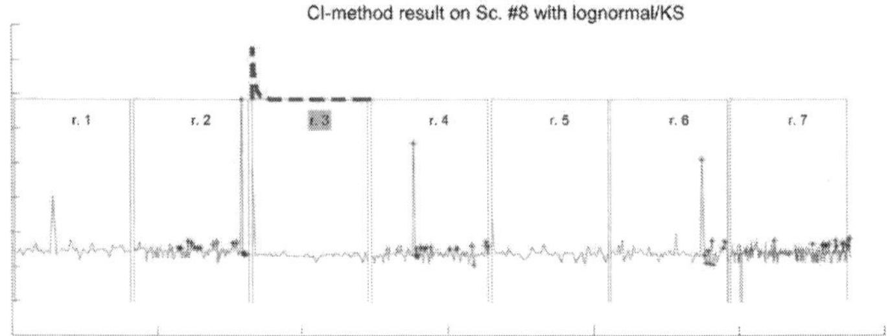

Cl-method result on Sc. #9 with lognormal/KS

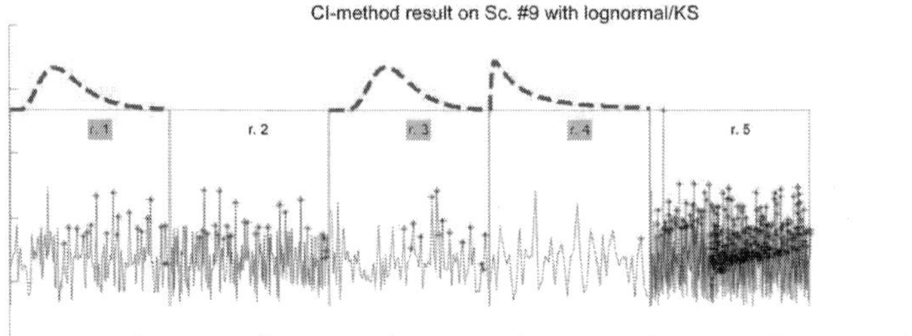

Cl-method result on Sc. #13 with lognormal/KS

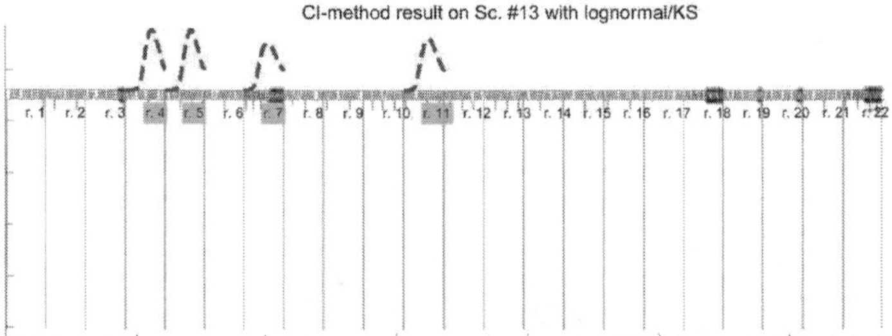

Cl-method result on Sc. #16 with lognormal/KS

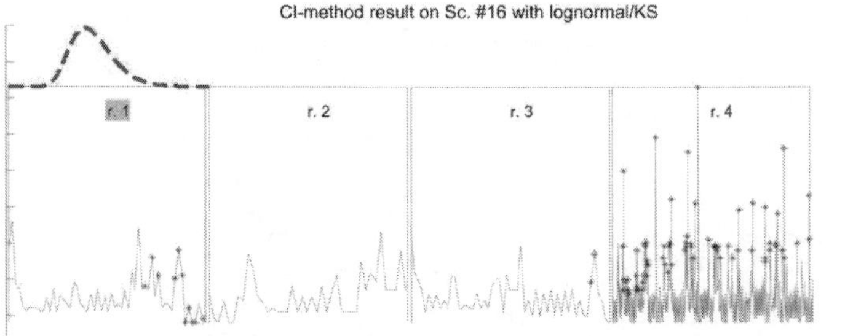

CI-method result on Sc. #17 with lognormal/CHI

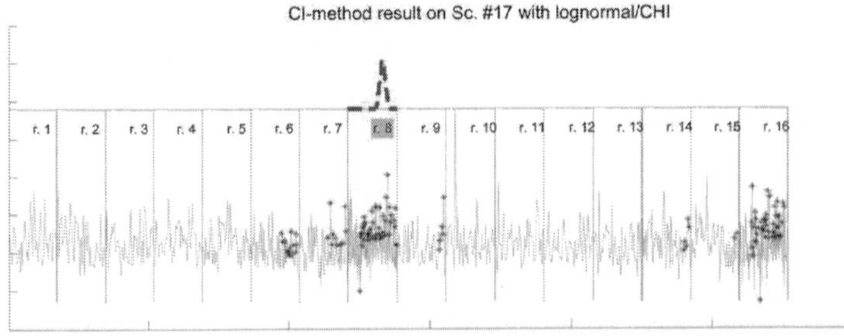

CI-method result on Sc. #17 with lognormal/KS

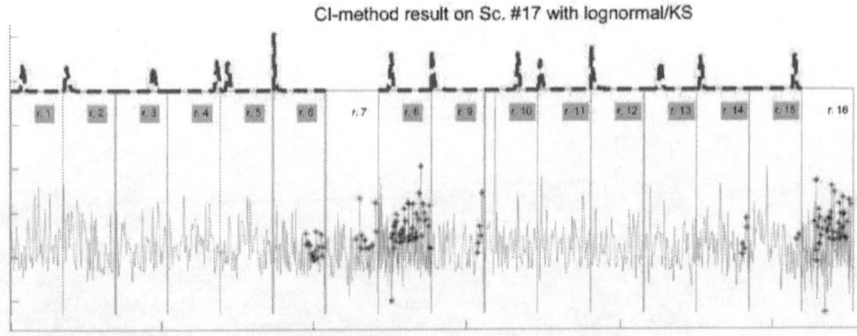

CI-method result on Sc. #17 with weibull/KS

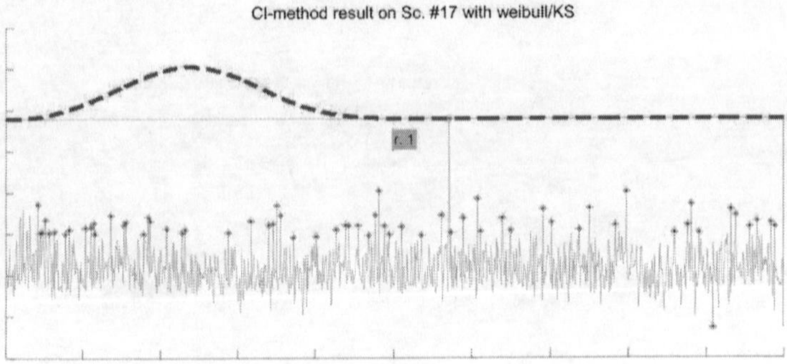

CI-method result on Sc. #18 with gamma/KS

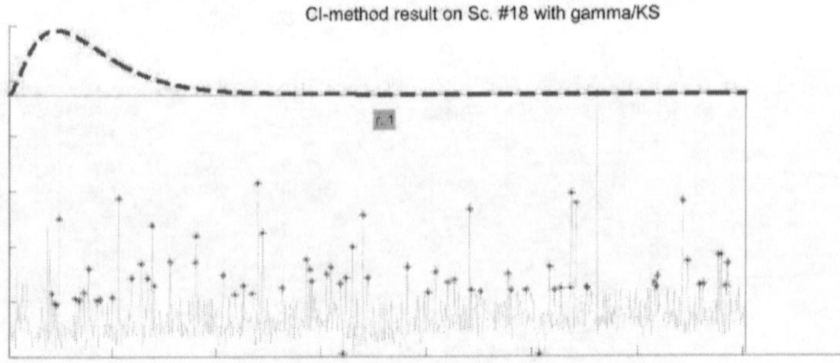

CI-method result on Sc. #18 with lognormal/KS

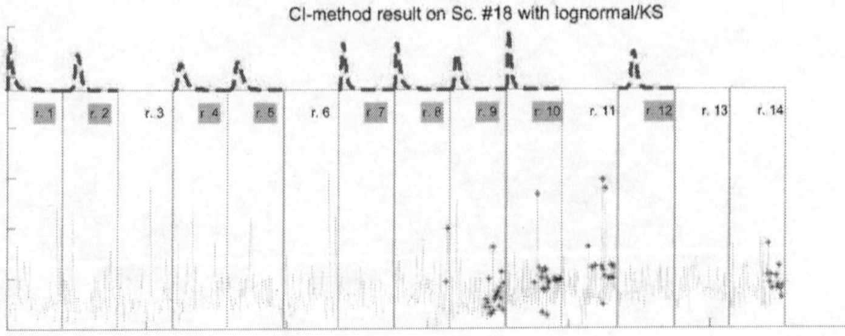

Figure 7. Plots of the detection of outliers, bursts and regime changes by our method of confidence intervals (CI) with different parametrical distributions. Black asterisks correspond to the detected bursts and outliers. All the models successfully found for the regimes detected by the algorithm (red lines over the green-marked segments) are depicted with the same aspect ratio in order to provide a fair comparison of their shapes. The delays below them have been stretched accordingly.

The results are shown in Table 8. Notice how the pattern already found with the confidence interval method, and even with unprocessed scenarios or split delays, appears here again: the lognormal model is the most successful one (although in this case all the others have a very good behaviour as well), both in the number of scenarios explained successfully and in the goodness of the models obtained, *i.e.*, the *p*-values.

4.4. Additional Analysis of Processed Delays

Apart from assessing the fitting of parametrical and non-parametrical models with hypothesis tests, we have also analysed in this work, with other statistical tools, the modeling results when using the confidence interval method: Quantile-Quantile plots and Probability Difference Graphs. They complete our statistical study of the problem, providing additional conclusions that are explained in the following:

Quantile-Quantile (Q-Q) plot

The quantile-quantile (Q-Q) plot is a graph of the input (observed) data values against the theoretical (fitted) distribution quantiles. Both axes of this graph are in units of the input data set. Q-Q plots can be used to determine, qualitatively, the goodness-of-fit of the models, but also other interesting questions such as how heavy their right tails are, which has consequences when used for predicting future time delays: a model that has heavier tail (greater area in the right tail) than the data that it is modeling will report a smaller probability of closing the sensory loop before the given time requirement than actually exists. That will produce pesimistic predictions, *i.e.*, longer expected delays, which can be useful in teleoperation [53]: in the case that there is no model good enough for representing the data tightly at all quantiles, it should be better to be pessimistic, *i.e.*, to work with the model that has heavier tail.

Table 8. Results obtained with the near-optimal change detection algorithm. The items marked in bold have the best p-value of their corresponding row, *i.e.*, they are the best models found for that scenario.

	Exponential (*p*-Value)	Erlang (*p*-Value)	Gamma (*p*-Value)	Lognormal (*p*-Value)	Weibull (*p*-Value)
χ^2	-	-	-	-	-
KS	**#1A(14;0.08)**	#1A(15;0.07)	#1A(12;0.04)	#1A(11;0.07)	#1A(15;0.07)
	#2B1(4;0.06)	#2B1(6;0.11)	#2B1(10;0.06)	**#2B1(2;0.15)**	#2B1(14;0.12)
	#3A/B2(11;0.04)	#3A/B2(11;0.04)	#3A/B2(8;0.03)	**#3A/B2(2;0.12)**	#3A/B2(14;0.06)
	#4B2(6;0.06)	**#4B2(8;0.24)**	#4B2(4;0.07)	#4B2(4;0.10)	#4B2(4;0.12)
	#5B2(3;0.51)	#5B2(4;0.20)	#5B2(2;0.04)	#5B2(0.03)	#5B2(5;0.21)
	#6B2(10;0.08)	**#6B2(10;0.19)**	#6B2(11;0.07)	#6B2(10;0.05)	#6B2(13;0.09)
	#7B2(10;0.03)	#7B2(12;0.06)	#7B2(10;0.10)	**#7B2(7;0.20)**	#7B2(7;0.08)
	#8C(12;0.12)	#8C(13;0.12)	#8C(13;0.14)	**#8C(12;0.168)**	#8C(9;0.167)
	#9A(22;0.04)	#9A(26;0.05)	#9A(21;0.04)	#9A(20;0.11)	**#9A(14;0.12)**
	#10B1(15;0.07)	#10B1(15;0.05)	#10B1(9;0.05)	**#10B1(11;0.09)**	#10B1(11;0.03)
	#11C(12;0.04)	**#11C(13;0.09)**	#11C(13;0.05)	#11C(14;0.06)	#11C(14;0.05)
	#12A(12;0.06)	**#12A(12;0.071)**	#12A(10;0.069)	#12A(11;0.05)	#12A(12;0.04)
	-	#13C(0.038)	#13C(10;0.039)	#13C(18;0.035)	**#13C(13;0.043)**
	#14C(0.03)	#14C(4;0.03)	#14C(7;0.03)	#14C(10;0.03)	**#14C(12;0.05)**
	#15A(12;0.04)	#15A(17;0.04)	#15A(13;0.04)	#15A(12;0.04)	**#15A(13;0.05)**
	#16B2(13;0.05)	**#16B2(14;0.13)**	#16B2(15;0.06)	#16B2(11;0.04)	#16B2(13;0.09)
	#17A(12;0.12)	#17A(13;0.14)	#17A(12;0.07)	**#17A(2;0.18)**	#17A(7;0.13)
	#18B1(9;0.03)	#18B1(12;0.09)	#18B1(3;0.05)	#18B1(2;0.06)	**#18B1(7;0.18)**
	85.5%	**78%**	**89.7%**	**92.9%**	**91.8%**

Figure 8 represents the Q-Q plot of all scenarios that we have found where different models have explained well the same segments of data (the only case when we can carry out a fair comparison). In general, all scenarios are better modeled in the lower quantiles, *i.e.*, the right tails of the distributions are not good models of the data, which is natural because there is less delays gathered in those tails. We can also observe that data are better modeled in scenarios #17 and #18 than in scenarios #4 and #5, but note that this is only a qualitative appreciation: all the depicted cases have passed the hypothesis test, therefore they are all statistically good as models of the data.

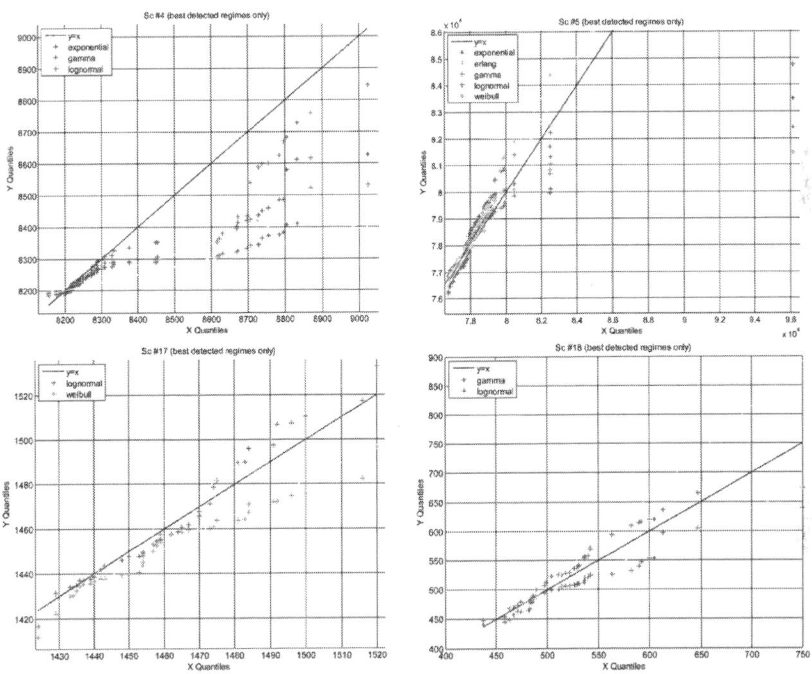

Figure 8. Q-Q plots for those regimes obtained by the confidence interval method where more than one parametrical model explains successfully the data obtaining their best *p*-values. The abscissas correspond to the data quantiles, while the ordinates are the models.

In that figure we see that Q-Q plots for scenarios #17 and #18 fit quite well the data. In contrast, scenarios #4 and #5 are arced, or "S" shaped, indicating worse models (although still within the acceptance of the hypothesis test). Also, all models there become more dispersed than the data in the highest quantiles, *i.e.*, in the right tails. In particular, the lognormal is the most heavy tailed model, and exponential/gamma the less heavy tailed. In these scenarios, where the models do not fit the data as well as in others, the lognormal distribution would predict less probability to close the sensory loop in a shorter time than the actual one, therefore being more conservative for doing control than the others. As explained before, being conservative is an interesting quality in our problem, since it facilitates to satisfy strict safety requirements with higher probability.

Probability Difference Graph (PDG)

The probability difference graph (Figure 9) is a plot of the difference between an empirical cumulative distribution function (cdf) and a theoretical cdf. This graph can be used to put in order the theoretical distributions in case we have to choose a model, since it yields a qualitative measure of distance between them. It also provides a general, qualitative indication of their goodness to fit the data: the closest to the zero horizontal, the better the fitting. Although all PDGs are for models that have been successful in modeling scenarios (according to the hypothesis test, reflected here in the small values of the curves in the ordinate axis), we can observe in the figure that in some cases the modeling has been better than in others. As happened in the Q-Q plots, scenario #4 has produced the worse fittings, which is reflected in the pronounced belly of all model curves around the middle of the delay values: if they are used for prediction they will be conservative in that area, which could be an interesting property as explained before. In that particular scenario, the gamma distribution seems slightly better than the other two, but their difference is rather small. In scenario #5 we obtain a pdf farther from zero when the erlang model is chosen as null hypothesis. In scenario #17 the lognormal has been better than the weibull. In scenario #18 the lognormal is also better for higher values of the delays, which is important in prediction.

5. CONCLUSIONS AND FUTURE WORK

This paper has shown how the time delay of the sensory data flow for a networked telerobot system, *i.e.*, the time that passes since a sensor request is issued in the client until the data is received and processed, has an interesting feature that allows us to model it in a minimalistic way: when abrupt changes in the delay sequence are properly detected, the remaining segments are mostly stationary and *iid.* Based on that result, we have presented a thorough statistical characterization of these sequences of delays, based on both parametrical and non-parametrical probability density functions that model marginal distributions of the delays. When applied to sequences of delays that have not been processed for detecting those abrupt changes, they do not show as particularly useful (with the exception of some non-parametricals), but we can improve those initial results firstly by splitting the delays into their natural additive components and then by applying a new method to detect non-linearities such as abrupt changes of regimes, outliers or bursts. The results reported in this paper can be summarised, qualitatively, as follows:

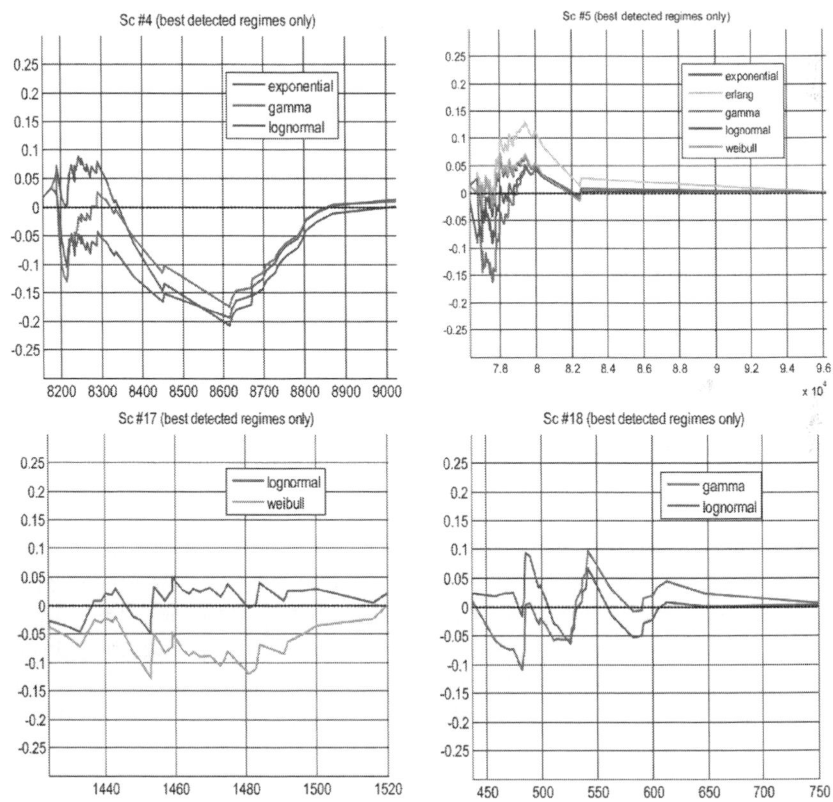

Figure 9. PDGs for those regimes obtained by the confidence interval method where more than one parametrical model explain successfully the data obtaining their best p-values.

- Unprocessed scenarios can only be reasonably well modelled with non-parametrical distributions. However, their use must be carefully considered, since they are more computational involved and prone to overfitting, *i.e.*, they model too well the already gathered delays at the expense of not modeling so well the delays to come just afterwards.

- Even when scenarios are unprocessed, the lognormal model performs better or equal than other parametrical ones. The kernel model is clearly better than the others in the case of non-parametrical models.

- If splitting the delays into three natural components (server + network + client delays) is possible, the performance of the lognormal model increases appreciably, while other parametrical models do not do so well. The kernel model also increases remarkably its performance with respect to other non-parametrical models, reaching near 70% of scenarios being explained. Therefore, kernel model could be the choice in that situation (again considering carefully the overfitting problem).

- Our simple confidence interval method for detecting abrupt changes in the delay sequence (only applicable to parametrical models) increases the performance of the lognormal as in the case of splitting the delays, but, in addition, it widens the applicability of the model, being able to explain scenarios of classes A, B and C (unlike in the splitting delays case). When there are relevant limitations in the computational cost available to execute on-line the change detection procedure, this method along with the lognormal should be the choice in general applications of the marginal modeling approach.

- As shown in the Q-Q plots, when using the confidence interval method, exponential and gamma models do not provide as good results as the lognormal, and in complicated scenarios that cannot be modeled well anyway, the latter could be a reasonable choice since it provides more conservative (pessimistic) estimates of the delays, which can serve to guarantee better the time requirements of the system.

- The chi-squared hypothesis test is not as good to detect whether a model explains well the data as the Kolmogorov-Smirnov test, mainly due to its sensitivity to short samples.

In this paper we have also identified intuitive categories of scenarios that cover most existing combinations of features that can appear in these sequences of delays (classes A, B and C). Just to link very briefly this classification to the actual results of our analyses, according to Section 3.3, when the scenarios are not processed for change detection, some scenarios from class B can be explained by parametrical models, which agrees with the fact that no abrupt changes with long lasting different regimes are present (Table 4); however the same models are not able to explain the rest of scenarios, which leads to a not very good utility of the unprocessing approach (this does not change even when delays are split into three additive components—network, server, client—, as shown in Table 5, which demonstrates that the features of the total delay signal are present in all of them, although the number of scenarios of class B that can be explained now is greater). Table 7 shows the modeling results when abrupt change detection is included through our confidence interval algorithm: then, classes B1 and B2 can be explained, as well as class A because the algorithm detects the accumulation of smooth trend fluctuation during certain time as an abrupt change at the end of that time, making the total fluctuation in each of these parts of the signal much smaller than in the original unprocessed scenario. Class C is more difficult to model, but the lognormal is able to explain some of those scenarios as well, including a largely deterministic one (scenario #13), included in our study to test the limits of stochasticity in our approach.

In the future we plan to analyse further this characterization problem (in particular, to devise new computationally efficient change detection algorithms that approach better the results of the near-optimal procedure explained in Section 4.3), and also to use these results to implement on-line dynamic estimators of the time delay, which will be the basis for a control of the sensor density that allows us to control in turn the networked telerobot in real-time as optimally as possible. In such environment, we will also develop methods for using more than one modeling procedure concurrently, aimed at guaranteeing the system performance in the minority of cases where the *iid* assumption cannot be satisfied.

ACKNOWLEDGMENTS

This work is supported by the Junta de Andalucía and the European Union (FEDER) through the research project P08-TIC-04282. We also wish to thank the anonymous reviewers of this paper for their insightful comments, which have served to make the original manuscript clearer, better structured and more precise, and Raúl Ruiz-Sarmiento, Eduardo Carletti and Ángel Martínez-Tenor for providing the data of some of the scenarios.

CONFLICTS OF INTEREST

The authors declare no conflict of interest.

REFERENCES

1. Jia, S.; Hada, Y.; Gang, Y.; Takase, K. Distributed Telecare Robotic Systems Using CORBA as a Communication Architecture. Proceedings of the IEEE International Conference on Robotics & Automation, Washington, DC, USA, 11–15 May 2002.

2. Cavusoglu, M.C.; Williams, W.; Tendick, F.; Sastry, S.S. Robotics for telesurgery: Second generation berkeley/UCSF laparoscopic telesurgical workstation and looking towards the future applications. *Ind. Robot* **2003**, *30*, 22–29.

3. Whitcomb, L. Underwater Robotics: Out of the Research Laboratory and into the Field. Proceedings of the IEEE International Conference on Robotics & Automation, Washington, DC, USA, 24–28 April 2000.

4. Tsumaki, Y.; Goshozono, T.; Abe, K.; Uchiyama, M.; Koeppe, R.; Hirzinger, G. Verification of an Advanced Space Teleoperation System Using Internet. Proceedings of the IEEE International Conference on Intelligent Robot Systems, Takamatsu, Japan, 30 October–5 November 2000.

5. Siciliano, B.; Khatib, O. *Springer Handbook of Robotics*; Springer: Berlin, Germany, 2008.

6. Burns, A.; Wellings, A. *Real-Time Systems and Programming Languages: Ada, Real-Time Java and C/Real-Time POSIX*; Addison-Wesley: Wokingham, UK, 2009.

7. Wang, Y.; Vuran, M.C.; Goddard, S. Cross-layer analysis of the End-to-End delay distribution in wireless sensor networks. *IEEE/ACM Trans. Netw* **2012**, *20*, 305–318.

8. Wang, X.; Schulzrinne, H. Comparison of adaptive internet multimedia applications. *IEICE Trans. Commun. E82-B* **1999**, 806–818.

9. Liu, P.X.; Meng, M.Q.H.; Gu, J.; Yang, S.X.; Hu, C. Control and Data Transmission for Internet Robotics. Proceedings of the IEEE International Conference on Robotics & Automation, Taipei, Taiwan, 14–19 September 2003.

10. Alt, G.H.; Lages, W.F. Networked Robot Control with Delay Compensation. Proceedings of the Fifth Real-Time Linux Workshop, Valencia, Spain, 9–11 November 2003.

11. Imer, O.C.; Yksel, S.; Basar, T. Optimal control of LTI systems over unreliable communication links. *Automatica* **2006**,*42*, 1429–1439.

12. Kim, W.; Ji, K.; Ambike, A. Networked real-time control strategy dealing with stochastic time delays and packet losses. *J. Dyn. Syst. Meas. Control* **2006**, *128*, 681–685.

13. Tipsuwan, Y.; Chow, M. Control methodologies in networked control systems. *J. Control Pract.* **2003**, *11*, 1099–1111.

14. Gibbon, J.F.; Little, T.D.C. The use of network delay estimation for multimedia data retrieval. *IEEE J. Sel. Areas Commun.* **1996**, *14*, 1376–1387.

15. YouTube. How YouTube Works—Computerphile. Available online: http://www.youtube.com/watch?v=OqQk7kLuaK4(accessed on 2 December 2013).

16. Yensen, T.; Lariviere, J.P.; Lambadaris, I.; Goubran, R.A. HMM delay prediction technique for VoIP. *IEEE Transa. Multimed.* **2003**, *5*, 444–457.

17. Lozoya Gámez, R.C.; Martí, P.; Velasco, M.; Fuertes, J.M. *Wireless Network Delay Estimation for Time-Sensitive Applications*, Research Report ESAII-RR-06-12. 2006. Available online: http:// esaii. upc.edu/people/pmarti/nde_06.pdf(accessed on 25 October 2013).

18. Deng, K.; Moore, A.W.; Nechyba, M.C. Learning to Recognize Time Series: Combining ARMA Models with Memory-Based Learning. Proceedings of the IEEE International Symposium on Computational Intelligence in Robotics and Automation, Monterey, California, USA, 10–11 July 1997.

19. Feng, H.; Shu, Y. Study on Network Traffic Prediction Techniques. Proceedings of the International Conference on Wireless Communications, Networking and Mobile Computing, Wuham, China, September 23–26 2005.

20. Li, A.; He, S.; Qin, Z. Real-Time Segmenting Time Series Data. In *Web Technologies and Applications. APWeb 2003*; Zhou, X., Zhang, Y., Orlowska, M.E., Eds.; Springer-Verlag: Berlin Heidelberg, Germany, 2003; Lecture Notes in Computer Science 2642; pp. 178–186.

21. De Buen Remiro, V. Modelación Masiva X-ARIMA. Available online: https://www.tol-project.org/svn/tolp/OfficialTolArchiveNetwork/ArimaTools/doc/Modelacion%20X-ARIMA%20masiva/Modelacion%20X-ARIMA%20masiva.pdf(accessed on 25 October 2013).

22. Gago-Benítez, A.; Fernández-Madrigal, J.A. Log-Normal Modeling of Non-Linear Sequences of Delays with Classical Change Detection Methods. Proceedings of the International Workshop on Applied Probability (IWAP'2012), Jerusalem, Israel, 11–14 June 2012.

23. Gago-Benítez, A.; Fernández-Madrigal, J.A.; Cruz-Martín, A. Log-logistic modelling of sensory flow delays in networked telerobots. *IEEE Sens.* **2013**, *13*, 2944–2953.

24. González-Jiménez, J.; Galindo, C.; Ruiz-Sarmiento, J.R. Technical Improvements of the Giraffe Telepresence Robot based on User's Evaluation. Proceedings of the 21st IEEE International Symposium on Robot and Human Interactive Communication (2012 IEEE RO-MAN), Paris, France, 9–13 September 2012.

25. Submarine Cable Map 2013. Available online: http://submarine-cable-map-2013.telegeography.com (accessed on 2 December 2013).

26. Giraff. Available online: http://www.giraff.org (accessed on 2 December 2013).

27. González, J.; Galindo, C.; Blanco, J.L.; Fernández-Madrigal, J.A.; Arévalo, V.; Moreno, F.A. SANCHO, a Fair Host Robot. A Description. Proceedings of the IEEE International Conference on Mechatronics, Málaga, Spain, 14–17 April 2009.

28. Surveyor Corporation. Available online: http://www.surveyor.com (accessed on 2 December 2013).

29. Arduino Homepage. Available online: http://arduino.cc (accessed on 2 December 2013).

30. Fernández-Madrigal, J.A.; Galindo, C.; González, J.; Cruz-Martín, E.; Cruz-Martín, A. A software engineering approach for the development of heterogeneous robotic applications. *J. Robot. Comput.-Integr. Manuf.* **2008**, *24*, 150–166.

31. Darie, C.; Brinzarea, B.; Chereches-Tosa, F.; Bucicia, M. *AJAX and PHP: Building Responsive Web Applications*; Packt Publishing: Birmingham, UK, 2006.

32. Song, D.; Goldberg, K.; Chong, N.Y. Networked Telerobots. In *Handbook of Robotics*; Siciliano, B., Khatib, O., Eds.; Springer: Berlin, Germany, 2008.

33. Goldberg, K.; Siegwart, R. Beyond Webcams. In *An Introduction to Online Robots*; The MIT Press: Boston, MA, USA, 2002.

34. Internet Devices. Available online: http://ford.ieor.berkeley.edu/ir/ (accessed on 2 December 2013).

35. Box, G.; Jenkins, G. *Time Series Analysis: Forecasting and Control*; Holden-Day: San Francisco, CA, USA, 1976.

36. D'Agostino, R.B.; Stephens, M.A. *Goodness-of-fit Techniques*; Marcel Dekker, Inc.: New York, NY, USA, 1986.

37. Jacobsson, K.; Hjalmarsson, H.; Niels Möller, N.; Johansson, K.H. *Round-Trip Time Estimation in Communication Networks using Adaptive Kalman Filtering*; Reglermöte 2004: Gothenburg, Sweden, 2004.

38. Cao, J.; Cleveland, W.S.; Lin, D.; Sun, D.X. Internet Traffic Tends Toward Poisson and Independent as the Load Increases. In

NonLinear Estimation and Classification; Springer: New York, NY, USA, 2002.

39. Walpole, R.E.; Myers, R.H.; Myers, S.L.; Ye, K.E. *Probability and Statistics for Engineers and Scientists*, 9th ed.; Pearson: Cranbury, NJ, USA, 2011.

40. Antoniou, I.; Ivanov, V.V.; Zrelov, P.V. On the log-normal distribution of network traffic. *J. Phys. D: Nonlinear Phenom.* **2002**, *167*, 72–85.

41. Wasserman, L. *All of Statistics: A Concise Course in Statistical Inference*; Springer: New York, NY, USA, 2003.

42. House, D.H. Spline Curves. Available online: http:// people.cs.clemson.edu/~dhouse/courses/405/notes/splines.pdf(accessed on 19 September 2013).

43. The MathWorks, Inc. Curve Fitting Toolbox 3.0. Available online: http://www.mathworks.com/products/curvefitting(accessed on 19 September 2013).

44. Barnett, V.; Lewis, T. *Outliers in Statistical Data*; John Wiley & Sons: San Francisco, California, USA, 1994.

45. Ben-Gal, I. Outlier Detection. In *Data Mining and Knowledge Discovery Handbook: A Complete Guide for Practitioners and Researchers*; Kluwer Academic Publishers: Berlin, Germany, 2005; pp. 131–146.

46. Ng, R.T.; Han, J. Efficient and Effective Clustering Methods for Spatial Data Mining. Proceedings of the Very Large Data Bases Conference, Zurich, Switzerland, 11–15 September 1994.

47. Acuna, E.; Rodriguez, C.A. Meta Analysis Study of Outlier Detection Methods in Classification. Technical paper, Department of Mathematics, University of Puerto Rico at Mayaguez. Available online: http://academic.uprm.edu/eacuna/paperout.pdf (accessed on 25 October 2013).

48. Lu, C.; Chen, D.; Kou, Y. Algorithms for Spatial Outlier Detection. Proceedings of the 3rd IEEE International Conference on Data-Mining (ICDM'03), Melbourne, FL, USA, 19–22 November 2003.

49. Knorr, E.; Ng, R. Algorithms for Mining Distance-Based Outliers in Large Datasets. Proceedings of the 24th International Conference Very Large Data Bases (VLDB), New York, NY, USA, 24–27 August 1998.

50. Knorr, E.; Ng, R.; Tucakov, V. Distance-based outliers: Algorithms and applications. *VLDB J.* **2000**, *8*, 237–253.

51. Bay, S.D.; Schwabacher, M. Mining Distance-Based Outliers in Near Linear Time with Randomization and a Simple Pruning Rule. Proceedings of the 9th ACM-SIGKDD Conference on Knowledge Discovery and Data Mining, Washington, DC, USA, 24–27 August 2003.

52. Cohen, A.; Whitten, B. Estimation in the three-parameter lognormal distribution. *J. Am. Stat. Assoc.* **1980**, *75*, 399–404.

53. Andreu, D.; Fraisse, P.; Roqueta, V.; Zapata, R. Internet Enhanced Teleoperation toward a Remote Supervised Delay Regulator. Proceedings of the IEEE International Conference on Industrial Technology, Maribor, Slovenia, 10–12 December 2003.

2

A NOVEL TECHNIQUE FOR DETECTION OF TIME DELAY SWITCH ATTACK ON LOAD FREQUENCY CONTROL

Arman Sargolzaei[1], Kang K. Yen[1], Mohamed N. Abdelghani[2], Abolfazl Mehbodniya[3], Saman Sargolzaei[4]

[1]Department of Electrical and Computer Engineering, Florida International University, Miami, FL, USA

[2]Department of Mathematics and Statistics, University of Alberta, Edmonton, Canada

[3]Graduate School of Engineering-Sendai, Tohoku University, Sendai, Japan

[4]Department of Electrical and Computer Engineering, Wentworth Institute of Technology, Boston, MA, USA

ABSTRACT

In this paper, we focus on the estimation of time delays caused by adversaries in the sensing loop (SL). Based on the literature review, time delay switch (TDS) attacks could make any control system, in particular a power control system, unstable. Therefore, future smart grids will have to use advanced methods to provide better situational awareness of power grid states keeping smart grids reliable and safe from TDS attacks. Here, we introduce a simple method for preventing time delay switch attack on networked control systems. The method relies on an estimator that will estimate and track time delays introduced by an adversary. Knowing the maximum tolerable time delay of the plant's optimal controller for which the plant remains stable, a time-delay detector issues an alarm signal when the estimated time delay is larger than the minimum one and directs the system to alarm state. In an alarm state, the plant operates under the control of an emergency controller that is local to the plant and remains in this mode until the networked control system state is restored. This method is an inexpensive and simple way to guarantee that an industrial control system remains stable and secure.

KEYWORDS

Time Delay Switch Attack, Load Frequency Control, Detection and Estimation, Emergency Controller

1. INTRODUCTION

Modern power grids rely on telecommunication technologies for control and monitoring, in a way to improve efficiency and reliability distribution. However, their reliance on computers and multi-purpose networks makes them vulnerable to cyber-attacks [1] -[3] . Recently, the investigation of attack methods on industrial control systems and their countermeasures have been the focus of many academic, industries, and governments. Here, we will investigate one type of attacks on a power control system recently proposed by Sargolzaei et al. [4] -[6] that is time delay switch attack and we provide a simple detection/estimation technique for time delays caused by an adversary. State estimation is one of the pivotal features of future power systems [7] . According to [8] , improving the resiliency and reliability of smart distribution networks could be obtained via deploying renewable resources and demand side management programs. As it has been mentioned in [9] , demand response can ameliorate the load curve by motivating smart customers utilizing smart communication infrastructure and smart meters.

The controller design for systems with time delay is one of the interests of many researchers [10] -[12] but they have not studied that time delay can be injected to the system as an attack. Time delays injected by a hacker in a control system would, in general, destabilize the system or cause inefficiency in performance of the system. This is a new attack in the context of power systems (e.g., load frequency control (LFC)) on networked control systems (NCS) [13] and is named time-delay-switch attack or TDS for short [4] . To avoid possible damage of TDS attacks, systems and controllers must be redesigned in a way to detect and correct for variable time delays.

Time delays exist in power systems, in the sensing and control loops. The traditional controllers of power systems are designed based on current information being available and ignoring time delays even if they present. However, power grids technologies are continuously being improved by introducing new telecommunication technologies for monitoring to improve efficiency, reliability and sustainability of supply and distribution. For example, the introduction of a wide area measurement system (WAMS) provides synchronized near real-time measurements in phase measurement units (PMUs). WAMS which are used for stability analysis of power systems can also be used for designing more robust controllers. Nevertheless, time delays are present in PMUs measurements as a result of natural transmission lines [14] .

Several studies have considered the problem of stability of power systems with time delays [10] [11] . The impact of time delays on the power system controllers was discussed in [15] -[19] . In [20] , authors studied the effects of delays on the small signal stability of power systems. In [15] and [16] , methods to eliminate oscillations that resulted from time-delayed feedback control were proposed. Authors in [21] presented a wide-area control system for damping generator oscillations. Using phasor measurements, but with delays, a controller was proposed in [14] and small signal stability of the power system was considered. A feedback controller designed for power systems with delayed states was proposed in [22] . This controller deals with the combined effects of the instantaneous as well as delayed states using the quadratic Lyapunov function for systems with delays. Additional studies on power systems with delay can be found in [2] [3] [10] -[12] [23] -[25] and the references therein. However, most of these studies considered either the construction of controllers that were robust to time delays or controllers that used offline estimate of time delays. As far as we know, there are no control methods that perform online estimation of dynamic time delays and real time control of power systems. Furthermore, few studies [26] [27] considered control of power systems with time delays introduced by adversary.

In this paper, we will describe a simple yet effective method to address a TDS attack on the observed states of a controlled system. Our method utilizes a time delay estimator, a communication protocol to alarm for time delay switch attack, a buffer to store the history of controller commands and an optimal controller to stabilize or track a reference signal and a local to the plant emergency controller to stabilize the plant if large time delays are detected. For now, we will only deal with LTI systems in state feedback.

2. METHOD

All control methods developed in the past compensate for time delays either rely on controller robust to a maximum time delay, off-line estimates of time delays or approximation of time delayed signals [26] [27] . Here, we develop an alternative time delay estimator which is simpler and will help with systems under time delay attack. In this paper, we propose a general method for control of systems under TDS attack. We have developed this method for continuous linear time-invariant systems. We have implemented our models and control strategies in MATLAB to demonstrate the performance with simulation.

The proposed method is shown in Figure 1. Its basic elements are: plant model, time delay detector, emergency controller and controller. The controller can either be a PID controller or an optimal feedback controller. The emergency controller's job is to stabilize the system in case of an attack on the communication lines of the distant controller. The time delay detector estimates the time delay in the communication channel from telemetered data. If the delay is larger than a maximum allowed time delay it sends alarm signals to the controller and emergency controller. If an alarm signal is received by the emergency controller it will begin to operate in a way to stabilize the plant to a desired state while the distant controller stops to operate until the delay is corrected.

Suppose the system we are dealing with is linear time invariant (LTI) or can be approximated in a region of interest by a LTI system,

$$\dot{x}(t) = Ax(t) + Bu(t), \tag{1}$$

where x and u are state and control vectors, respectively. Matrices A and B are constant matrices with suitable dimensions.

Then, the solution is given by

$$x(t) = e^{At}x_0 + \int_0^t e^{A(t-s)}Bu(s)ds \tag{2}$$

with time delay τ, either a time-delay switch attack or a natural delay, the solution of Equation (2) becomes

$$x(t-\tau) = e^{A(t-\tau)}x_0 + \int_0^{t-\tau} e^{A(t-\tau-s)}Bu(s)ds \tag{3}$$

Let us write the solution $x(t)$ at the time t in terms of the solution with the time delay τ; we have

$$\begin{aligned} x(t) &= e^{At}x_0 + e^{A\tau}\int_0^{t-\tau} e^{A(t-s)}e^{-A\tau}Bu(s)ds + \int_{t-\tau}^{t} e^{A(t-s)}Bu(s)ds \\ &= e^{At}x_0 + e^{A\tau}\left[x(t-\tau) - e^{A(t-\tau)}x_0\right] + \int_{t-\tau}^{t} e^{A(t-s)}Bu(s)ds, \end{aligned} \tag{4}$$

In general, the time delay τ is an unknown variable. Let's assume that τ is slowly varying, compared to the changes in u and x, and $\hat{\tau}$ is our estimate of the time delay τ. Then, $\varepsilon = \hat{\tau} - \tau$ is the estimation error in the time delay. The predicted state $\hat{x}(t)$ of the system based on the estimate of time delay $\hat{\tau}$ is given by

$$\hat{x}(t) = e^{At}x_0 + e^{A\hat{\tau}}\left[\hat{x}(t-\hat{\tau}) - e^{A(t-\hat{\tau})}x_0\right] + \int_{t-\hat{\tau}}^{t} e^{A(t-s)}Bu(s)ds \tag{5}$$

where $\hat{x}(t-\hat{\tau})$ is the estimate of the delayed state given the estimate of the delay $\hat{\tau}$ (i.e. a simulated signal).

It should be noted that, $x(t-\tau)$ is what we actually measures and deliver to the plant model. So, at every instance of time, the variables $\hat{x}(t)$, $\hat{x}(t-\hat{\tau})$, $u(t)$, A, B and $x(t-\tau)$ are known to the controller and the plant model. On the other hand, the current state $x(t)$ and the time delay τ are unknown. It's essential that the plant model estimates state $x(t)$ accurately. Because of the delay, an accurate enough estimation of $x(t)$ requires a good estimate of the delay τ. We will show how to estimate the delay τ.

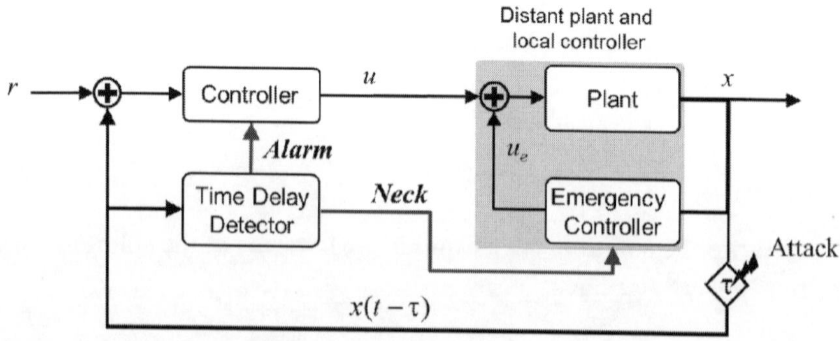

Figure 1. Block diagram of the proposed control system.

The estimation error in states can be described by $e_m(t)=x(t)-\hat{x}(t)$ and with delay it is given by

$$e_m(t;\tau,\hat{\tau})=x(t-\tau)-\hat{x}(t-\hat{\tau}) \qquad .(6)$$

Then the idea is to estimate $\hat{\tau}$ overtime as fast as possible to minimize the error $e_m(t;\tau,\hat{\tau})$. Using the gradient descent method, we set

$$\frac{d\hat{\tau}}{dt}=-\eta\frac{\partial v}{\partial\hat{\tau}}, \qquad (7)$$

where η is the learning parameter and select $v=\frac{1}{2}e_m^2$. With some manipulation, we have

$$\begin{aligned}\frac{\mathrm{d}\hat{\tau}}{\mathrm{d}t} &= -\eta e_m \frac{\partial e_m}{\partial \hat{\tau}} = -\eta e_m \frac{\partial\left[x(t-\tau)-\hat{x}(t-\hat{\tau})\right]}{\partial \hat{\tau}} \\ &= \eta e_m \frac{\partial \hat{x}(t-\hat{\tau})}{\partial \hat{\tau}} = \eta e_m \frac{\partial}{\partial \hat{\tau}}\left[\mathrm{e}^{A(t-\hat{\tau})}x_0 + \int_0^{t-\hat{\tau}} \mathrm{e}^{A(t-\hat{\tau}-s)}Bu(s)\mathrm{d}s\right] \\ &= \eta e_m \frac{\partial}{\partial \hat{\tau}}\left[\int_0^{t-\hat{\tau}} \mathrm{e}^{A(t-\hat{\tau}-s)}Bu(s)\mathrm{d}s\right] - \eta e_m A\mathrm{e}^{A(t-\hat{\tau})}x_0 \\ &= -\eta e_m\left[Bu(t-\hat{\tau}) - \mathrm{e}^{A(t-\hat{\tau})}Bu(0) - A\mathrm{e}^{A(t-\hat{\tau})}x_0\right]\end{aligned} \tag{8}$$

Assuming that, $u(0)=0$, which is reasonable for initial time, we arrive at

$$\frac{\mathrm{d}\hat{\tau}}{\mathrm{d}t} = -\eta e_m Bu(t-\hat{\tau}) - A\mathrm{e}^{A(t-\hat{\tau})}x_0, \qquad 0 \le \hat{\tau} \le t \tag{9}$$

Equation (9) is the one we will use to estimate the time delay τ. However, there are practical issues that we need to consider. Computing machines have finite memory and temporal resolution. Therefore, we are unable to implement Equation (9) without discrete approximation and boundedness assumptions. To guarantee stability of calculations and limit our memory usage we have to add the following condition, $\tau < \tau_{\max}$. This condition will allow us to construct a finite buffer to store the history of $u(t)$ from t to $t-\tau_{\max}$.

After designing the time delay estimator, we turn our attention to the controller and emergency controller. The controller can either be a PID controller or an optimal controller depending on the requirements of the application. Equation (10) is the PID controller and Equation (11) is the optimal controller,

$$u(t) = K_P e(t) + K_D \frac{\mathrm{d}e}{\mathrm{d}t}(t) + K_I \int_0^t e(s)\mathrm{d}s, \tag{10}$$

$$u(t) = Ke(t). \tag{11}$$

where the error is $e(t)=r(t)-x(t)$. Either controller, i.e., the PID or the optimal controller, can be designed in a way to be robust to some maximum time-delay τ_{stable}. The emergency controller sits close to the plant and operates only in case of emergency, for example, an attack on the communication line between the plant and controller. The emergency controller could either be a PID controller or optimal controller with the goal to stabilize the system to a particular reference trajectory r_E.

Suppose there is a time delay attack on the system with delay τ; the time delay estimator will estimate the delay $\hat{\tau}$. The cyber-attack detector will use the time delay estimate to perform the following function

$$D = \begin{cases} 1 & \hat{\tau} > c\tau_{\text{stable}} \\ 0 & \text{otherwise} \end{cases}, \tag{12}$$

where c is a constant between 0 and 1. In case D = 1 an alarm signal is sent to the controller to shut it down and a negative acknowledgement is sent to the emergency controller to stabilize the plant. The control strategy is shown in Figure 1.

We have implemented this method in MATLAB and verified its performance using the load frequency control (LFC) of a two-area distributed power system. In the next section we present and discuss the simulation results.

3. RESULT

We focus on the LFC system where the controller's function is to regulate the states of a networked power plant. The multi interconnect LFC dynamic system description can be found in [4] . Here we focus on two-area power system with the attack model as

$$\begin{cases} \dot{X}(t) = AX(t) + BU(t) + D\Delta P_l \\ X(0) = X_0 \end{cases} . \tag{13}$$

The optimal feedback controller is given by

$$U(t) = -K\hat{X}(t) \tag{14}$$

and the new state after the attack can be modeled by

$$\hat{X}(t)=X(t-\tau), \tag{15}$$

where $\tau=[t_{d1},t_{d2},\cdots,t_{dN}]^{\mathrm{T}}$ are different/random time-delays and are positive values. When $t_{d1},t_{d2},\cdots,t_{dN}$ are

all zero, the system is in its normal operation. An adversary can get access to the communication link and inject a delay attack on the line to direct the system to abnormal operations.

In (13), $X(t)=[x_1(t) \;\; x_2(t)]^{\mathrm{T}}$ denotes the states in the first and second power areas. Then the state vector in the ith power area is defined as

$$x_i(t)=[\Delta f^i(t) \;\; \Delta P_g^i(t) \;\; \Delta P_{tu}^i(t) \;\; \Delta P_{pf}^i(t) \;\; \Lambda^i(t)]^{\mathrm{T}} \qquad i=1,2, \tag{16}$$

where $\Delta f^i(t)$, $\Delta P_g^i(t)$, $\Delta P_{tu}^i(t)$, $\Delta P_{pf}^i(t)$ and $\Lambda^i(t)$ are frequency deviation, power deviation of the generator, position value of the turbine, tie-line power flow, and control error on the ith power area, respectively [19] . The control error of the ith power area is expressed as

$$\Lambda^i(t)=\int_0^t \beta_i \Delta f^i(s)\,\mathrm{d}s, \tag{17}$$

where β_i denotes the frequency bias factor.

In the dynamic model of the two-area LFC (13):

$$A=\begin{bmatrix} A_{11} & A_{12} \\ A_{21} & A_{22} \end{bmatrix}, \tag{18}$$

$$B=\mathrm{diag}\left\{\left[B_1^{\mathrm{T}} \;\; B_2^{\mathrm{T}}\right]^{\mathrm{T}}\right\}, \tag{19}$$

$$D=\mathrm{diag}\left\{\left[D_1^{\mathrm{T}} \;\; D_2^{\mathrm{T}}\right]^{\mathrm{T}}\right\}, \tag{20}$$

and A_{ii}, A_{ij}, B_i and D_i are represented by

$$A_{ii} = \begin{bmatrix} -\frac{\mu_i}{J_i} & \frac{1}{J_i} & 0 & -\frac{1}{J_i} & 0 \\ 0 & -\frac{1}{T_{tu\,i}} & \frac{1}{T_{tu\,i}} & 0 & 0 \\ -\frac{1}{\omega_i T_{g\,i}} & 0 & -\frac{1}{T_{g\,i}} & 0 & 0 \\ \sum_{\substack{i \neq j \\ j=1}}^{N} 2\pi T_{ij} & 0 & 0 & 0 & 0 \\ \beta_i & 0 & 0 & 0 & 1 \end{bmatrix} \quad (21)$$

$$B_i = \begin{bmatrix} 0 & 0 & \frac{1}{T_{g\,i}} & 0 & 0 \end{bmatrix}^T \quad (22)$$

$$A_{ij} = \begin{bmatrix} 0 & 0 & 0 & 0 & 0 \\ 0 & 0 & 0 & 0 & 0 \\ 0 & 0 & 0 & 0 & 0 \\ -2\pi T_{ij} & 0 & 0 & 0 & 0 \\ 0 & 0 & 0 & 0 & 0 \end{bmatrix} \quad (23)$$

and

$$D_i = \begin{bmatrix} -\frac{1}{J_i} & 0 & 0 & 0 & 0 \end{bmatrix}^T \quad (24)$$

where i and j can only be values one and two, J_i, ω_i, μ_i, $T_{g\,i}$ and $T_{tu\,i}$ are the generator moment of inertia, the speed-droop coefficient, generator damping coefficient, the governor time constant, the turbine time constant in the ith power area and T_{ij} is the stiffness constant between the ith and the jth power areas, respectively.

The analysis starts with the design of an optimal controller for the LFC in the normal operation (i.e., with no attack). Consider the system model described by (13) with the performance index described by

$$J=\frac{1}{2}\int_{0}^{t_f}\left\{X^{\mathrm{T}}(t)QX(t)+U^{\mathrm{T}}(t)RU(t)\right\}\mathrm{d}t, \tag{25}$$

where matrix $Q\in R^{n\times n}$ is positive semi-definite and $R\in R^{m\times m}$ is positive definite. Then, the optimal control problem is to obtain optimal control $U^{*}(t)$ that minimizes the performance index (25), subject to the dynamic of the system with no time-delay in its states.

Simulation studies have been conducted to evaluate the effects of TDS attacks on the dynamics of the system and detect TDS attacks to direct the controller to emergency state. By solving the Riccati matrix equation we obtain the close loop control law in the form of state feedback. For the simulations we have used discrete linear-quadratic regulator design from continuous cost function called "lqrd" function in MATLAB 2013a. Table 1 shows parameter values used in this process. We also set both ΔP_l^1 and ΔP_l^2 to zero.

In our simulation, the total simulation time is 50 second and the sampling time is 0.01 s as its common in industrial applications. To show accuracy of our proposed TDS attack detector/tracker, we didn't send negative acknowledgment to the local controller for the first simulation to only track the injected time delay to the system. The simulation has been done for three different scenarios: 1) single TDS attack on one power area, 2) Simultaneous TDS attack on both power areas, 3) Complex varied TDS attack on both power areas.

1) Single TDS attack to one power area: Here, we considered that adversary attacks the third state of the first power area at time of 2 second for a delay value of 3 seconds and the time delay is increased to 4.5 seconds at time of 7 seconds. Figure 2 shows that the detector accurately tracked and monitored the TDS attack or any natural delay on the system.

2) Simultaneous TDS attack to both power areas: In this scenario, we attack the third state of both power areas. TDS attacks 1 and 2 have been injected to the first and the second power areas respectfully. TDS attack 1 is started at time 2 seconds for value 3 seconds and increased at time 8 seconds to the value of 4.5 seconds. Also TDS attack 2 is started simultaneously with TDS attack 1 for the value of 1.5 seconds and increased to the value of 6 seconds. Figure 3 shows the result.

Table 1. Parameter values for a two-area power system controller design.

Parameter	Value	Parameter	Value
J_1	10	ω_1	0.05
μ_1	1.5	T_{g1}	0.12 s
T_{tu1}	0.2 s	T_{tu2}	0.45 s
T_{12}	0.198 pu/rad	T_{21}	0.198 pu/rad
J_2	12	ω_2	0.05
μ_2	1	T_{g2}	0.18 s
R	100 I	Q_f	0
Q	100 I	t_f	∞
β_1	21.5	β_2	21

3) Complex varied TDS attack to both power areas: In the last scenario we injected the TDS attack at different time with different time-delay values. We assume that an adversary injects TDS attack to feedback lines of both power areas. In our simulation, an attacker starts TDS attack to the second power area (third state) at time 1 second for $t_{d3} = 5$ s and then increases it to $t_{d3} = 10$ s at 20 second. Furthermore, adversary starts to attack the first power area (third state), at 1 second with $t_{d8} = 3$ s and increased it to $t_{d8} = 4.5$ s at 30 second. Figure 4 and Figure 5 show TDS attack tracking real time. Figure 4 shows TDS attack on the third state of the first power area andFigure 5 shows the attack on the same state of the second power area.

For the second part of our simulation we enabled the emergency controller to show the effect of our proposed technique for overcoming the TDS attack. We assume that an adversary injects TDS attack to the feedback lines of both power areas. In the simulation, a TDS attack was applied to the second power area (the third state) at time of 1 second for $t_{d3} = 5$ s and then increases it to $t_{d3} = 10$ s at 20 second. Furthermore, adversary starts to attack the first power area (the third state), at 1 second with $t_{d8} = 3$ s and increased it to $t_{d8} = 4.5$ s at 30 second. We also set the $\tau_{stable} = 0.4$ s based on stability analysis of LFC system in [4] . We send the negative acknowledgment to emergency controller in the case that detected value of time-delay is more than the maximum allowed time-delay. Figure 6 and Figure 7 show the third state of the first and the second power areas under attack with traditional optimal controller (TOC) and the proposed control technique (PCT) respectively. As it's clear on the result, the simulated attack makes the system unstable. With the proposed technique, we could overcome TDS attack on the simulated system.

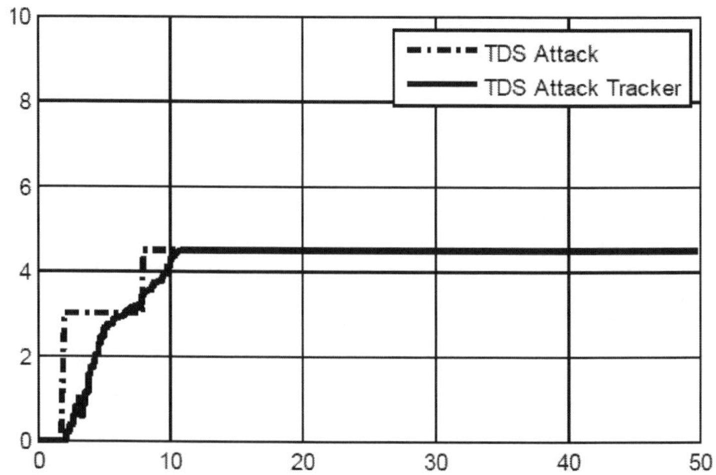

Figure 2. TDS attack detection and tracking for third state of first power area.

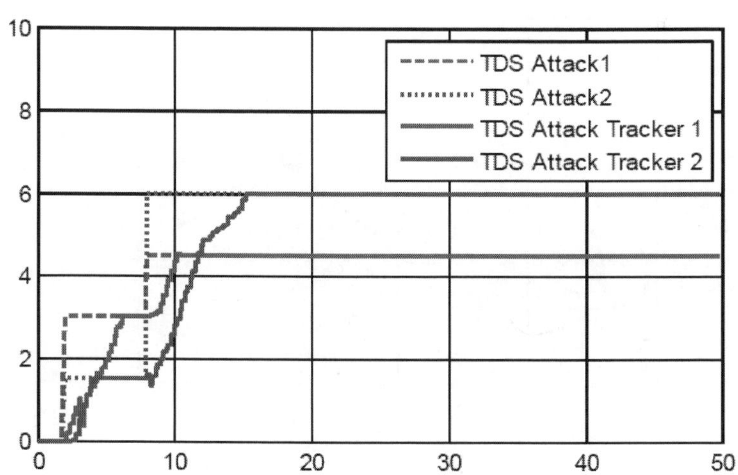

Figure 3. TDS attack detection and tracking for simultaneous TDS attack on both power areas.

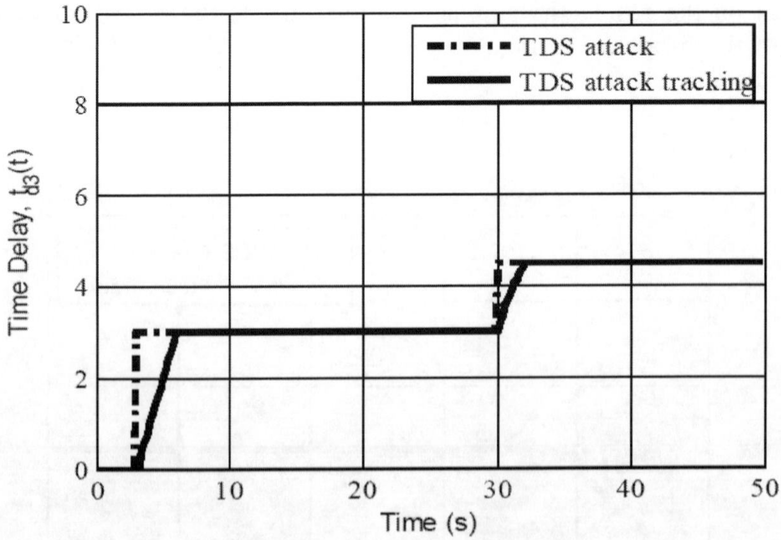

Figure 4. TDS attack detection and tracking for third state of first power area.

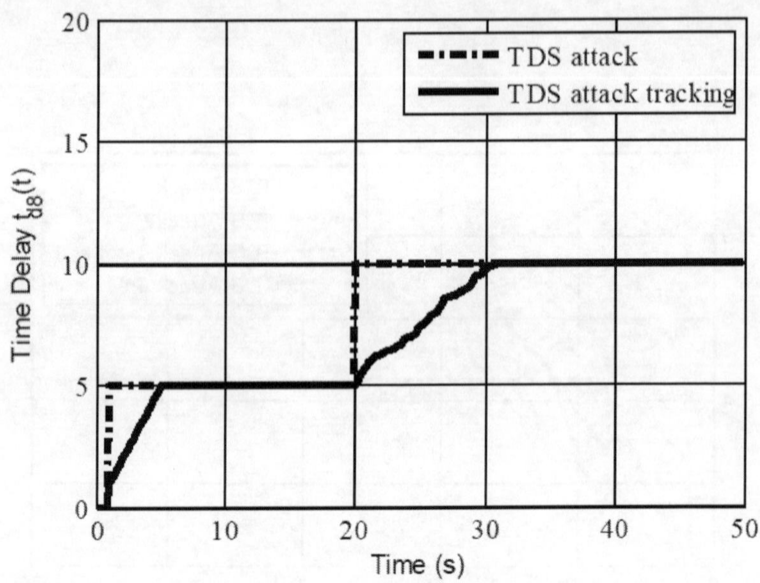

Figure 5. TDS attack detection and tracking for third state of second power area.

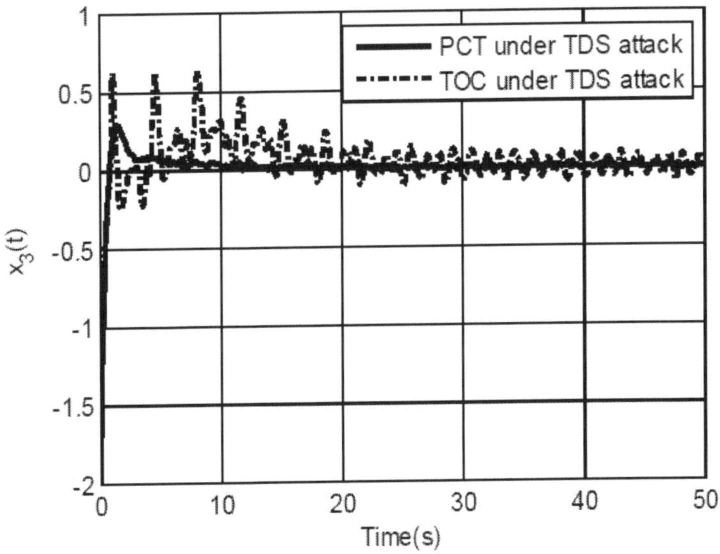

Figure 6. Value position of the turbine system under TDS attack for the first power-area control system.

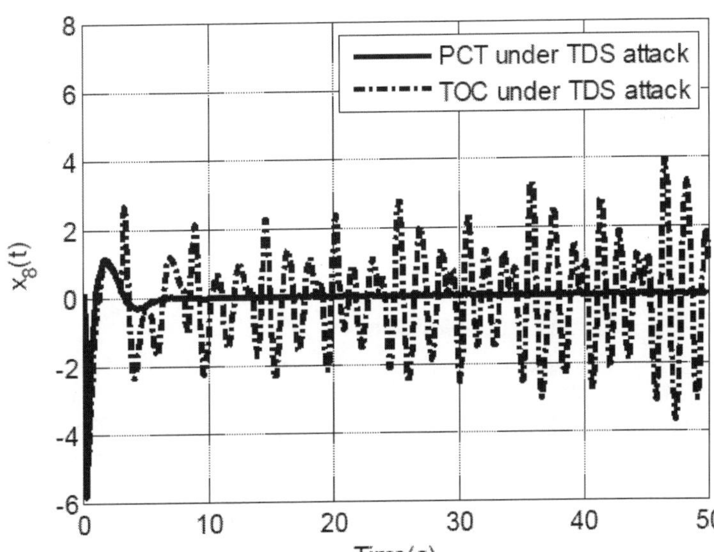

Figure 7. Value position of the turbine system under TDS attack for the second power-area control system.

4. CONCLUSION

In summary, we have demonstrated a simple method for estimating and detecting time delay switch attack on a networked control system. The method relies on a time delay estimator that estimates and tracks time-delays introduced by an adversary. With knowledge of the maximum time delay of the control system, for which the plant remains stable and secure, the time-delay detector compares the estimated time delay to the maximum allowed time delay and issues an alarm signal when the estimated time delay is larger than this value. It also directs the system to an alarm state. In an alarm state, the plant is under the control of the emergency controller, which is local to the plant. The plant remains in this mode until the networked control system state is restored and time-delay switch is eliminated. We think that this method is simple and an inexpensive way to assure that an industrial control system remains stable and secure.

ACKNOWLEDGEMENTS

The authors would like to thank the Resilient, Autonomous Networked Control Systems (RANCS) group for support during publishing this paper.

REFERENCES

1. Ericsson, G.N. (2010) Cyber Security and Power System Communication—Essential Parts of a Smart Grid Infrastructure. IEEE Transactions on Power Delivery, 25, 1501-1507.

2. Zhang, C.-K., Jiang, L., Wu, Q.H., He, Y. and Wu, M. (2013) Delay-Dependent Robust Load Frequency Control for Time Delay Power Systems. IEEE Transactions on Power Systems, 28, 2192-2201.

3. Yuan Y.L., Li Z.Y. and Ren, K. (2011) Modeling Load Redistribution Attacks in Power Systems. IEEE Transactions on Smart Grid, 2, 382-390.

4. Sargolzaei, A., Yen, K. and Abdelghani, M.N. (2014) Time-Delay Switch Attack on Load Frequency Control in Smart Grid. Journal of Advanced Communication Technologies, 5, 55-64.

5. Sargolzaei, A., Yen, K. and Abdelghani, M.N. (2014) Delayed Inputs Attack on Load Frequency Control in Smart Grid. Innovative Smart Grid Technologies Conference (ISGT), 2014 IEEE PES, Washington DC, 19-22 February 2014, 1-5.

6. Sargolzaei, A., Yen, K.K. and Abdelghani, M.N. (2014) Control of Nonlinear Heartbeat Models under Time-Delay-Switched Feedback Using Emotional Learning Control. International Journal on Recent Trends in Engineering & Technology, 10, 2.

7. Amini, M.H., Sarwat, A.I., Iyengar, S.S. and Guvenc, I. (2014) Determination of the Minimum-Variance Unbiased Estimator for DC Power-Flow Estimation. 40th IEEE Industrial Electronics Conference (IECON 2014), Dallas, 29 October-1 November 2014, 114-118.

8. Amini, M.H., Nabi, B. and Haghifam, M.-R. (2013) Load Management Using Multi-Agent Systems in Smart Distribution Network. IEEE PES General Meeting 2013, Vancouver, 21-25 July 2013, 1-5

9. Kamyab, F., Amini, M. H., Sheykhha, S., Hasanpour, M. and Jalali, M.M. (2015) Demand Response Program in Smart Grid Using Supply Function Bidding Mechanism. IEEE Transactions on Smart Grid, 25 May 2015, 1949-3053.

10. Li, Y., Tong, S.C. and Li, Y.M. (2012) Observer-Based Adaptive Fuzzy Backstepping Control for Strict-Feedback Stochastic Nonlinear Systems with Time Delays. International Journal of Innovative Computing, Information and Control, 8, 8103-8114.

11. Benzaouia, A., Ouladsine, M., Naamane, A. and Ananou, B. (2012) Fault Detection for Uncertain Delayed Switching Discrete-Time Systems. International Journal of Innovative Computing, Information and Control, 8, 8049-8062.

12. Sargolzaei, A., Yen, K., Noei, S. and Ramezanpour, H. (2013) Assessment of He's Homotopy Perturbation Method for Optimal Control of Linear Time-Delay Systems. Applied Mathematical Sciences, 7, 349-361.

13. Benéitez-Péerez, H., Benéitez-Péerez, A. and Ortega-Arjona, J. (2012) Networked Control Systems Design Considering Scheduling Restrictions and Local Faults. International Journal of Innovative Computing, Information and Control, 8, 8515-8526.

14. Dotta, D., Silva, A.S. and Decker, I.C. (2009) Wide-Area Measurements-Based Two-Level Control Design Considering Signal Transmission Delay. IEEE Transactions on Power Systems, 24, 208-216.

15. Kamwa, I., Grondin, R. and Hebert, Y. (2001) Wide-Area Measurement Based Stabilizing Control of Large Power Systems—A Decentralized/Hierarchical Approach. IEEE Transactions on Power Systems, 16, 136-153.

16. Wu, H.X., Tsakalis, K.S. and Heydt, G.T. (2004) Evaluation of Time Delay Effects to Wide-Area Power System Stabilizer Design. IEEE Transactions on Power Systems, 19, 1935-1941.

17. Jiang, Q., Zou, Z. and Cao, Y. (2005) Wide-Area TCSC Controller Design in Consideration of Feedback Signals' Time Delays. IEEE Power Engineering Society General Meeting, 2, 1676-1680.

18. Saad, M.S., Hassouneh, M.A., Abed, E.H. and Edris, A.A. (2005) Delaying Instability and Voltage Collapse in Power Systems Using SVCs with Washout filter-Aided Feedback. American Control Conference, 6, 4357-4362.

19. Chaudhuri, B., Majumder, R. and Pal, B. (2005) Wide-Area Measurement Based Stabilizing Control of Power System Considering Signal Transmission Delay. IEEE Power Engineering Society General Meeting, 2, 1447-1450.

20. Milano, F. and Anghel, M. (2012) Impact of Time Delays on Power System Stability. IEEE Transactions on Circuits and Systems I: Regular Papers, 59, 889-900.

21. Ray, S. and Venayagamoorthy, G.K. (2008) Real-Time Implementation of a Measurement-Based Adaptive Wide-Area Control System Considering Communication Delays. IET Generation, Transmission & Distribution, 2, 62-70.

22. Alrifai, M.T., Zribi, M., Rayan, M. and Mahmoud, M.S. (2013) On the Control of Time Delay Power Systems. International Journal of Innovative Computing, Information and Control, 9, 769-792.

23. Yang, Q.Y., An, D. and Yu, W. (2013) On Time Desynchronization Attack against IEEE 1588 Protocol in Power Grid Systems. 2013 IEEE Energytech, Cleveland, 21-23 May 2013, 1-5.

24. Schenato, L. (2008) Optimal Estimation in Networked Control Systems Subject to Random Delay and Packet Drop. IEEE Transactions on Automatic Control, 53, 1311-1317.

25. Mahmoud, M.S. (2000) Robust Control and Filtering for Time-Delay Systems. Marcel Dekker Inc., New York.

26. Tan, Y.H. (2004) Time-Varying Time-Delay Estimation for Nonlinear Systems Using Neural Networks. International Journal of Applied Mathematics and Computer Science, 14, 63-68.

27. Li, C.M. and Xiao, J. (2006) Adaptive Delay Estimation and Control of Networked Control Systems. International Symposium on Communications and Information Technologies, 2006, ISCIT'06, Bangkok, 18-20 October 2006, 707-710.

3

STABILIZATION OF UNKNOWN NONLINEAR DISCRETE-TIME DELAY SYSTEMS BASED ON NEURAL NETWORK

Vinay Kumar Deolia[1], Shubhi Purwar[2], Tripti Nath Sharma1

[1]Department of Electronics & Communications Engineering, G. L. A. University, Mathura, India

[2]Department of Electrical Engineering, Motilal Nehru National Institute of Technology (MNNIT), Allahabad, India

ABSTRACT

This paper discusses about the stabilization of unknown nonlinear discrete-time fixed state delay systems. The unknown system nonlinearity is approximated by Chebyshev neural network (CNN), and weight update law is presented for approximating the system nonlinearity. Using appropriate Lyapunov-Krasovskii functional the stability of the nonlinear system is ensured by the solution of linear matrix inequalities. Finally, a relevant example is given to illustrate the effectiveness of the proposed control scheme.

KEYWORDS

Discrete-Time Delay Nonlinear Systems; Lyapunov Stability; Linear Matrix Inequality (LMI)

1. INTRODUCTION

Over the past few decades, time-delay systems have drawn much attention from researchers throughout the world. This is due to their important role in many practical systems. A number of research results concerning time-delay systems exist in literature [1-6] and the references therein. As time-delay is a main source of instability and poor performance, considerable attention has been given for analysis and synthesis of such systems.

Since most of the physical systems are in continuous time, a considerable amount of attention has been payed to stability and control of continuous-time linear systems with delay [7-11]. Delay independent and, less conservative, delay dependent sufficient stability conditions in terms of Riccati or linear matrix inequalities (LMIs) have been derived by using Lyapunov-Krasovskii functional (LKF) or Lyapunov-Razumikhin functions. For continuous-time systems with uncertain, non-small delay a new construction of the LKF has been introduced in [12]. To a nominal LKF, which is appropriate to the nominal system (with nominal delays), terms are added which correspond to the perturbed system and which vanish when the delay perturbations approaches to zero [3]. Discretetime delay systems have drawn less attention as compared to continuous-time delay systems. For linear discrete-time delay systems, the contribution in [13] is worth mentioning. In [13], the delay involved in the system is removed at the cost of increased order of the system. However, the systems involving large delays, the proposed scheme in [13] is invariably leads to large scale systems. Furthermore, for systems with unknown or time-varying delays the proposed technique in [13] cannot be applied. In [14], delay dependent and independent conditions have been derived for determining the asymptotic stability of discrete-time systems with uncertain delay, time varying delay and norm-bounded uncertainties. Robust stability and guaranteed cost control problem is solved in [3], for discrete-time delay systems. The work on delay-dependent robust stabilization of uncertain discrete-time state-delayed systems is proposed in [15].

For continuous-time delay nonlinear systems the work on adaptive neural network control with unknown time delays is reported in [16]. Adaptive neural control of nonlinear time-delay systems with unknown virtual control coefficients is proposed in [17]. In [18] work is presented on adaptive neural control for a class of nonlinearly parametric time-delay system. Backstepping control for a class of delayed nonlinear systems with input constraints is reported in [19]. Fuzzy approximation based adaptive control of strict-feedback nonlinear systems with time delays is shown in [20].

In [21] problem of feedback stabilization of nonlinear discrete-time systems with delays is explained. In this by using the Lyapunov-Razumikhin approach, general conditions for stabilizing the closed-loop system is derived. The stability analysis of discrete-time systems with time varying state-delay is shown in [22]. By defining a new Lyapunov functions and by making use of novel techniques to achieve delay dependence, several new conditions are obtained for the asymptotic stability of these systems. The problem of designing nonlinear observers for dynamic discrete-time systems with both constant and time-varying delay nonlinearities is addressed in [23]. In [23] the nonlinear system is assumed to verify the usual Lipschitz condition that permits to transform the nonlinear system into a linear time-delay system with structured uncertainties. An optimal control scheme for a class of discrete-time nonlinear systems with time delays in both state and control variables with respect to a quadratic performance index function using adaptive dynamic programming is presented in [24]. State feedback stabilization of discrete-time delay nonlinear system is reported in [25]. In [25] a simple LMI condition is obtained for asymptotic stabilization through an appropriate Lyapunov function and judicious mathematical manipulations. An explicit state feedback law is provided that may be seen as a generalization of the existing results.

The main significance of this paper is to guarantee the stabilization of the unknown single-input-single-output (SISO) delayed systems. The unknown nonlinear functions of the system are approximated through Chebyshev neural networks (CNNs). Through a suitable LyapunovKrasovskii functional, conditions are derived to guarantee the stabilization of the nonlinear system in terms of simple LMIs. The organization of this paper is as follows: The CNN structure is presented in Section 2. Problem formulation is given in Section 3. The stability analysis is detailed in Section 4. To show the effectiveness of proposed scheme simulation results are reported in Section 5. The note ends with conclusion in Section 6.

Notations: $\|\cdot\|$ denotes Euclidean norm, $\|\cdot\|_F$ implies Frobenius norm. The tr stands for trace of matrix. Upper subscript T is the transpose of matrix and the symmetric entries in a symmetric matrix are given by *.

2. CNN STRUCTURE

Artificial neural networks (ANNs) emerged as powerful learning technique to perform complex tasks in highly nonlinear dynamic environment. Some of the basic advantages of ANN models are: their ability to learn on the basis of optimization of an appropriate error function and their excellent performance for approximation of nonlinear functions. As an alternative to multilayer perceptron (MLP), radial basis function (RBF) networks have been considered, primarily because of its simple structure. The RBF networks can learn functions with local variations and discontinuities effectively and also possess universal approximation capability [26]. This network represents a function of interest by using members of family of compactly or locally supported basis functions, out of which radially-symmetric Gaussian functions, are found to be quite popular. A RBF network has been proposed for effective identification of nonlinear dynamic systems [27, 28]. In these networks, however, choosing an appropriate set of RBF centers for effective learning still remains a problem.

The major drawback of feed forward neural network such as a MLP trained with back propagation (BP) algorithm is that it requires a large amount of computation for learning. A single-layer functional link artificial neural network (FLANN) in which the need of hidden layer is eliminated by expanding the input pattern using Chebyshev polynomials. The main advantage of this network is that it requires much less computation as compared to a multilayer perceptron (MLP) [29].

The structure of Chebyshev neural network (CNN) is shown in **Figure 1** CNN is a functional link network (FLN) based on Chebyshev polynomials. There are two main parts in the architecture of CNN, namely, numerical transformation part and learning part. Numerical transformation deals with input to hidden layer by approximate transformable method. The transformation is a functional expansion (FE) of input pattern comprising of a finite set of Chebyshev polynomials. As a result the Chebyshev polynomials basis can be viewed as a new input vector. The learning part is a functional-link neural network based on Chebyshev polynomials [30-33]. The Chebyshev polynomials can be obtained by a recursive formula

$$T_{i+1}(x) = 2xT_i(x) - T_{i-1}(x), T_0(x) = 1 \quad (1)$$

where, $T_i(x)$ are Chebyshev polynomials, i is the order of polynomials chosen and here x is a scalar quantity. The different choices of $T_i(x)$ are x & 2x.

The output of single layer neural network is given by

$$\hat{g}(x) = \hat{w}^T \varphi \tag{2}$$

where, w are the weights and φ is the suitable basis function of neural network. Based on the approximation property of CNN [30-33], there exist ideal weights w, so that the function $g(x)$ to be approximated can be represented as

$$g(x) = w^T \varphi + \varepsilon \tag{3}$$

where, ε is the CNN functional reconstruction error vector and $\|\varepsilon\| \le \varepsilon_N$ is bounded.

Approximation of complex nonlinear systems becomes easier as CNN is a single layer neural network.

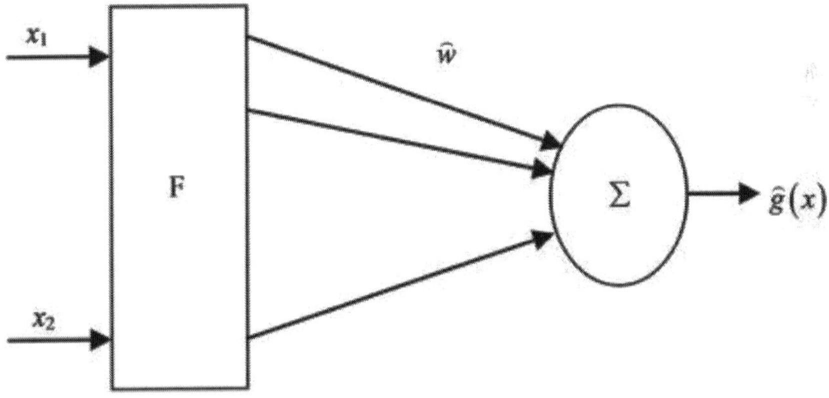

Figure 1. Chebyshev neural network.

3. PROBLEM FORMULATION

Consider the discrete-time nonlinear system with a fixed known delay [25]

$$x(k+1) = Ax(k) + A_d x(k-h) + g(x(k))u(k) \tag{4}$$

where, $x(k) \in R^n$ and $u(k) \in R^m$ denote the state and input vectors respectively at time instant k. A and A_d are constant matrices of appropriate dimensions. $g(x(k))$ is a unknown nonlinear function of a given system in (4), and h is a positive known number representing delay.

For the system (4) the stabilizing controller is chosen as

$$u(k) = -\frac{1}{\hat{g}(x(k))}\left[2Ax(k) + 2A_d x(k-h)\right] \tag{5}$$

where $\hat{g}(x(k))$ is the approximated value of unknown system function $\hat{g}(x(k))$. Let us assume $\hat{g}(x(k)) \geq 0$.

The objective of the current work is to ensure the stability and performance of the nonlinear system (4), using the controller (5) and appropriate Lyapunov-Krasovskii functional (11).

4. STABILITY ANALYSIS

For the stability analysis, the following assumptions are needed to proceed.

Assumption 1: (Bounded Ideal NN Weights): The ideal NN weights w are bounded so that $\|w_F\| \leq w_M$, with w_M a known bound. The symbol $\|\cdot\|_F$ denotes the Frobenius norm, i.e. given a matrix A, the Frobenius norm is given by [34],

$$\|A\|_F^2 = tr(A^T A)$$

Assumption 2: Let $\tilde{g}(x(k)) = G\hat{g}(x(k))$, where $G = G^T$ is a $n \times n$ symmetric matrix, and $\tilde{g}(x(k))$ and $\hat{g}(x(k))$ are the n-column vectors.

Equation (4) can be rewritten as

$$x(k+1) = Ax(k) + A_d x(k-h) + \tilde{g}(x(k))u(k) + \hat{g}(x(k))u(k) \quad (6)$$

where

$$\tilde{g}(x(k)) = g(x(k)) - \hat{g}(x(k)) \quad (7)$$

The main results are as follows:

Theorem 1:

Suppose there exist an $n \times n$ positive-definite matrix P, an $n \times n$ nonnegative-definite matrix Q, and $n \times n$ symmetric matrix G such that following LMIs holdsH1)

$$\begin{bmatrix} A^T PA - \boldsymbol{P} + \boldsymbol{Q} & A^T PA_d \\ * & A_d^T PA_d - \boldsymbol{Q} \end{bmatrix} > 0 \quad (8)$$

and H2)

$$\begin{bmatrix} A^T \xi A + A^T \zeta A & A^T \xi A_d + A^T \zeta A_d \\ * & A_d^T \xi A_d + A_d^T \zeta A_d \end{bmatrix} < 0 \quad (9)$$

where $\xi = PG$ and $\zeta = G^T PG$ then, the nonlinear discrete-time delay system (4) ensures the stability with the control law in (5) and weight tuning algorithm

$$\hat{w}(k+1)=\Big[\hat{w}(k)+\left\{x^T(k)\left[A^TPA+\boldsymbol{Q}-\boldsymbol{P}\right]x(k)\right\}^{\frac{1}{2}}$$
$$+\left\{x^T(k-h)\left[A_d^TPA_d-\boldsymbol{Q}\right]x(k-h)\right\}^{\frac{1}{2}}$$
$$+\left\{x^T(k-h)\left[A_d^TPA\right]x(k)\right\}^{\frac{1}{2}}$$
$$+\left\{x^T(k)\left[A^TPA_d\right]x(k-h)\right\}^{\frac{1}{2}}\Big]$$

.(10)

Proof:

Choose Lyapunov-Krasovskii functional,

$$V(k)=V_1(k)+V_2(k)+V_3(k) \tag{11}$$

where,

$$V_1(k)=x^T(k)Px(k) \tag{12}$$

$$V_2(k)=\sum_{i=k-h}^{k-1}x^T(i)Qx(i) \tag{13}$$

$$V_3(k)=tr\left(\tilde{w}^T(k)\tilde{w}(k)\right) \tag{14}$$

Substituting (12)-(14) in (11)

$$V(k)=x^T(k)Px(k)+\sum_{i=k-h}^{k-1}x^T(i)Qx(i)+tr\left(\tilde{w}^T(k)\tilde{w}(k)\right) \tag{15}$$

and

$$V(k+1)=x^T(k+1)Px(k+1)+\sum_{i=k+1-h}^{k}x^T(i)Qx(i)+tr\left(\tilde{w}^T(k+1)\tilde{w}(k+1)\right) \tag{16}$$

Since P is a positive-definite and Q is a nonnegative-definite, $V(k)$ is then positive-definite [25]. Therefore,

$$\Delta V(k) = V(k+1) - V(k) \tag{17}$$

Substituting (15) and (16) in (17),

$$\begin{aligned}\Delta V(k) &= x^T(k+1)Px(k+1) + \sum_{i=k+1-h}^{k} x^T(i)Qx(i) \\ &\quad + tr\left(\tilde{w}^T(k+1)\tilde{w}(k+1)\right) - x^T(k)Px(k) \\ &\quad - \sum_{i=k-h}^{k-1} x^T(i)Qx(i) - tr\left(\tilde{w}^T(k)\tilde{w}(k)\right)\end{aligned} \tag{18}$$

Substituting (6) in (18),

$$\begin{aligned}&\Delta V(k) \\ &= \left[Ax(k) + A_d x(k-h) + \tilde{g}(x(k))u(k)\right. \\ &\quad \left.+\hat{g}(x(k))u(k)\right]^T P\left[Ax(k) + A_d x(k-h)\right. \\ &\qquad \left.+\tilde{g}(x(k))u(k) + \hat{g}(x(k))u(k)\right] \\ &\quad + x^T(k)Qx(k) + tr\left(\tilde{w}^T(k+1)\tilde{w}(k+1)\right) - x^T(k)Px(k) \\ &\quad - x^T(k-\mathrm{h})Qx(k-\mathrm{h}) - tr\left(\tilde{w}^T(k)\tilde{w}(k)\right)\end{aligned} \tag{19}$$

Using Assumption 1 in (19) and further solving (19),

$$\begin{aligned}&V(k) \\ &= x^T(k)\left[A^T PA\right]x(k) + x^T(k)\left[A^T PA_d\right]x(k-h) \\ &\quad + x^T(k)A^T P\hat{g}(x(k))u(k) + x^T(k-h)\left[A_d^T PA\right]x(k) \\ &\quad + x^T(k-h)\left[A_d^T PA_d\right]x(k-h) \\ &\quad + x^T(k-h)A_d^T P\hat{g}(x(k))u(k) \\ &\quad + u^T(k)\hat{g}^T(x(k))PAx(k) \\ &\quad + u^T(k)\hat{g}^T(x(k))PA_d x(k-h) \\ &\quad + u^T(k)\hat{g}^T(x(k))P\hat{g}(x(k))u(k) \\ &\quad + Z(k) + x^T(k)Qx(k) - x^T(k)Px(k) \\ &\quad - x^T(k-h)Qx(k-h) + \left\|\tilde{w}(k+1)\right\|^2 - \left\|\tilde{w}(k)\right\|^2\end{aligned} \tag{20}$$

where,

$$
\begin{aligned}
& Z(k) \\
& = x^T(k) A^T P \tilde{g}(x(k)) u(k) + x^T(k-h) A_d^T P \tilde{g}(x(k)) u(k) \\
& + u^T(k) \tilde{g}^T(x(k)) P A x(k) + u^T(k) \tilde{g}^T(x(k)) P A_d x(k-h) \\
& + u^T(k) \tilde{g}^T(x(k)) P \tilde{g}(x(k)) u(k) \\
& + u^T(k) \tilde{g}^T(x(k)) P \hat{g}(x(k)) u(k) \\
& + u^T(k) \hat{g}^T(x(k)) P \tilde{g}(x(k)) u(k)
\end{aligned} \tag{21}
$$

In (20) collecting the terms together, substitute control law from (5), adding and subtracting some terms to make perfect square yields,

$$
\begin{aligned}
& \Delta V(k) \\
& = -\|\hat{w}(k+1)\|^2 \\
& \quad + \Bigg[\Bigg\{\Bigg\{ \hat{w}(k) + \left\{x^T(k)\left[A^T P A + Q - P\right]x(k)\right\}^{\frac{1}{2}} \\
& \qquad\quad + \left\{x^T(k-h)\left[A_d^T P A_d - Q\right]x(k-h)\right\}^{\frac{1}{2}} \Bigg\} \\
& \qquad + \left\{x^T(k-h)\left[A_d^T P A\right]x(k)\right\}^{\frac{1}{2}} \Bigg\} \\
& \quad + \left\{x^T(k)\left[A^T P A_d\right]x(k-h)\right\}^{\frac{1}{2}} \Bigg]^2 + S(k) + Z(k)
\end{aligned} \tag{22}
$$

where, $S(k)$ is given in (23).

From (22) the weight tuning law is obtained as (24),

$$
\begin{aligned}
S(k) = & -2\left\{\left\{\hat{w}(k)+\left\{x^{T}(k)\left[A^{T}PA+Q-P\right]x(k)\right\}^{\frac{1}{2}}+\left\{x^{T}(k-h)\left[A_{d}^{T}PA_{d}-Q\right]x(k-h)\right\}^{\frac{1}{2}}\right\}\right. \\
& \left.+\left\{x^{T}(k-h)\left[A_{d}^{T}PA\right]x(k)\right\}^{\frac{1}{2}}\right\}\left\{x^{T}(k)\left[A^{T}PA_{d}\right]x(k-h)\right\}^{\frac{1}{2}} \\
& -2\left\{\hat{w}(k)+\left\{x^{T}(k)\left[A^{T}PA+Q-P\right]x(k)\right\}^{\frac{1}{2}}+\left\{x^{T}(k-h)\left[A_{d}^{T}PA_{d}-Q\right]x(k-h)\right\}^{\frac{1}{2}}\right\}\left\{x^{T}(k-h)\left[A_{d}^{T}PA\right]x(k)\right\}^{\frac{1}{2}} \\
& -2\left\{\hat{w}(k)+\left\{x^{T}(k)\left[A^{T}PA+Q-P\right]x(k)\right\}^{\frac{1}{2}}\right\}\left\{x^{T}(k-h)\left[A_{d}^{T}PA_{d}-Q\right]x(k-h)\right\}^{\frac{1}{2}} \\
& -2\hat{w}(k)\left\{x^{T}(k)\left[A^{T}PA+Q-P\right]x(k)\right\}^{\frac{1}{2}}
\end{aligned}
\tag{23}
$$

$$
\begin{aligned}
\hat{w}(k+1) = & \left[\left\{\left\{\hat{w}(k)+\left\{x^{T}(k)\left[A^{T}PA+Q-P\right]x(k)\right\}^{\frac{1}{2}}+\left\{x^{T}(k-h)\left[A_{d}^{T}PA_{d}-Q\right]x(k-h)\right\}^{\frac{1}{2}}\right\}\right.\right. \\
& \left.\left.+\left\{x^{T}(k-h)\left[A_{d}^{T}PA\right]x(k)\right\}^{\frac{1}{2}}\right\}+\left\{x^{T}(k)\left[A^{T}PA_{d}\right]x(k-h)\right\}^{\frac{1}{2}}\right]
\end{aligned}
\tag{24}
$$

then (22) becomes,

$$
\Delta V(k) = S(k) + Z(k) \tag{25}
$$

Substituting (5), (21) and (23) in (25), gives (26).

Using Assumption 2, (26) is given by (27).

$$
\begin{aligned}
V(k) = & -2\left\{\left\{\hat{w}(k)+\left\{x^{T}(k)\left[A^{T}PA+Q-P\right]x(k)\right\}^{\frac{1}{2}}+\left\{x^{T}(k-h)\left[A_{d}^{T}PA_{d}-Q\right]x(k-h)\right\}^{\frac{1}{2}}\right\}\right. \\
& \left.+\left\{x^{T}(k-h)\left[A_{d}^{T}PA\right]x(k)\right\}^{\frac{1}{2}}\right\}\left\{x^{T}(k)\left[A^{T}PA_{d}\right]x(k-h)\right\}^{\frac{1}{2}} \\
& -2\left\{\hat{w}(k)+\left\{x^{T}(k)\left[A^{T}PA+Q-P\right]x(k)\right\}^{\frac{1}{2}}\right. \\
& \left.+\left\{x^{T}(k-h)\left[A_{d}^{T}PA_{d}-Q\right]x(k-h)\right\}^{\frac{1}{2}}\right\}\left\{x^{T}(k-h)\left[A_{d}^{T}PA\right]x(k)\right\}^{\frac{1}{2}} \\
& -2\left\{\hat{w}(k)+\left\{x^{T}(k)\left[A^{T}PA+Q-P\right]x(k)\right\}^{\frac{1}{2}}\right\}\left\{x^{T}(k-h)\left[A_{d}^{T}PA_{d}-Q\right]x(k-h)\right\}^{\frac{1}{2}} \\
& -2\hat{w}(k)\left\{x^{T}(k)\left[A^{T}PA+Q-P\right]x(k)\right\}^{\frac{1}{2}}+2x^{T}(k)A^{T}P\tilde{g}(x(k))\left\{-\frac{1}{\hat{g}(x(k))}\left[2Ax(k)+2A_{d}x(k-h)\right]\right\} \\
& +2x^{T}(k-h)A_{d}^{T}P\tilde{g}(x(k))\left\{-\frac{1}{\hat{g}(x(k))}\left[2Ax(k)+2A_{d}x(k-h)\right]\right\} \\
& +2\left\{-\frac{1}{\hat{g}(x(k))}\left[2Ax(k)+2A_{d}x(k-h)\right]\right\}^{T}\tilde{g}^{T}(x(k))P\hat{g}(x(k))\left\{-\frac{1}{\hat{g}(x(k))}\left[2Ax(k)+2A_{d}x(k-h)\right]\right\} \\
& +\left\{-\frac{1}{\hat{g}(x(k))}\left[2Ax(k)+2A_{d}x(k-h)\right]\right\}^{T}\tilde{g}^{T}(x(k))P\tilde{g}(x(k))\left\{-\frac{1}{\hat{g}(x(k))}\left[2Ax(k)+2A_{d}x(k-h)\right]\right\}
\end{aligned}
\tag{26}
$$

$$\begin{aligned}\Delta V(k) = & -2\hat{w}(k)\left\{x^T(k)\left[A^T PA_d\right]x(k-h)\right\}^{\frac{1}{2}} - 2\hat{w}(k)\left\{x^T(k-h)\left[A_d^T PA\right]x(k)\right\}^{\frac{1}{2}} \\ & -2\hat{w}(k)\left\{x^T(k-h)\left[A_d^T PA_d - Q\right]x(k-h)\right\}^{\frac{1}{2}} - 2\hat{w}(k)\left\{x^T(k)\left[A^T PA + Q - P\right]x(k)\right\}^{\frac{1}{2}} \\ & -2\left\{x^T(k)\left[A^T PA + Q - P\right]x(k)\right\}^{\frac{1}{2}}\left\{x^T(k)\left[A^T PA_d\right]x(k-h)\right\}^{\frac{1}{2}} \\ & -2\left\{x^T(k)\left[A^T PA + Q - P\right]x(k)\right\}^{\frac{1}{2}}\left\{x^T(k-h)\left[A_d^T PA\right]x(k)\right\}^{\frac{1}{2}} \\ & -2\left\{x^T(k-h)\left[A_d^T PA_d - Q\right]x(k-h)\right\}^{\frac{1}{2}}\left\{x^T(k)\left[A^T PA_d\right]x(k-h)\right\}^{\frac{1}{2}} \\ & -2\left\{x^T(k-h)\left[A_d^T PA_d - Q\right]x(k-h)\right\}^{\frac{1}{2}}\left\{x^T(k-h)\left[A_d^T PA\right]x(k)\right\}^{\frac{1}{2}} \\ & -2\left\{x^T(k-h)\left[A_d^T PA\right]x(k)\right\}^{\frac{1}{2}}\left\{x^T(k)\left[A^T PA_d\right]x(k-h)\right\}^{\frac{1}{2}} \\ & -2\left\{x^T(k)\left[A^T PA + Q - P\right]x(k)\right\}^{\frac{1}{2}}\left\{x^T(k-h)\left[A_d^T PA_d - Q\right]x(k-h)\right\}^{\frac{1}{2}} \\ & +x^T(k)\left[4A^T PGA + 4A^T G^T PGA\right]x(k) + x^T(k)\left[4A^T PGA_d + 4A^T G^T PGA_d\right]x(k-h) \\ & +x^T(k-h)\left[4A_d^T PGA + 4A_d^T G^T PGA\right]x(k) + x^T(k-h)\left[4A_d^T PGA_d + 4A_d^T G^T PGA_d\right]x(k-h)\end{aligned} \tag{27}$$

Let,

$$a = \left\{x^T(k)\left[A^T PA + Q - P\right]x(k)\right\}^{\frac{1}{2}}\left\{x^T(k)\left[A^T PA_d\right]x(k-h)\right\}^{\frac{1}{2}}$$
$$b = \left\{x^T(k-h)\left[A_d^T PA_d - Q\right]x(k-h)\right\}^{\frac{1}{2}}\left\{x^T(k-h)\left[A_d^T PA\right]x(k)\right\}^{\frac{1}{2}}$$
$$c = \left\{x^T(k)\left[A^T PA_d + Q - P\right]x(k)\right\}^{\frac{1}{2}}\left\{x^T(k-h)\left[A_d^T PA\right]x(k)\right\}^{\frac{1}{2}}$$
$$d = \left\{x^T(k-h)\left[A_d^T PA_d - Q\right]x(k-h)\right\}^{\frac{1}{2}}\left\{x^T(k)\left[A^T PA_d\right]x(k-h)\right\}^{\frac{1}{2}}$$

Manipulating the nonquadratic terms using the following inequality

$$\sqrt{ab} \le \frac{a+b}{2}$$

(which turns into equality if and only if a = b) we get (28).

In (28), $\Delta V(k)$ is guaranteed to remain negative as long as

$$\begin{bmatrix} A^T PA - P + Q & A^T PA_d \\ * & A_d^T PA_d - Q \end{bmatrix} > 0$$

and

$$\begin{bmatrix} A^T\xi A + A^T\zeta A & A^T\xi A_d + A^T\zeta A_d \\ * & A_d^T\xi A_d + A_d^T\zeta A_d \end{bmatrix} < 0$$

where $\xi = PG$ and $\zeta = G^T PG.$

Since in (28) the first four quadratic terms are negative, next four terms are satisfying LMI in (8), and remaining last four terms are satisfying LMI in (9). Therefore, we can conclude that the system (4) is stable with control law (5) and two LMIs in (8) and (9).

5. SIMULATION RESULTS

A numerical example is presented to demonstrate the performance and the effectiveness of the proposed scheme.

Consider the non-linear discrete-time delay system (4) with the following parameters [25]:

$$A = \begin{bmatrix} 0.8 & 0.4 \\ -0.14 & 0.1 \end{bmatrix}, A_d = \begin{bmatrix} 0.2 & 0 \\ 0 & 0.1 \end{bmatrix}$$

and

$$\begin{aligned}
\Delta V(k) \le -4\Bigg(& \left\{x^T(k)\left[A^T PA + Q - P\right]x(k)\right\}^{\frac{1}{2}} \left\{x^T(k)\left[A^T PA_d\right]x(k-h)\right\}^{\frac{1}{2}} \\
& \left\{x^T(k-h)\left[A_d^T PA_d - Q\right]x(k-h)\right\}^{\frac{1}{2}} \left\{x^T(k-h)\left[A_d^T PA\right]x(k)\right\}^{\frac{1}{2}}\Bigg)^{\frac{1}{2}} \\
& -4\Bigg(\left\{x^T(k)\left[A^T PA + Q - P\right]x(k)\right\}^{\frac{1}{2}} \left\{x^T(k-h)\left[A_d^T PA\right]x(k)\right\}^{\frac{1}{2}} \\
& \left\{x^T(k-h)\left[A_d^T PA_d - Q\right]x(k-h)\right\}^{\frac{1}{2}} \left\{x^T(k)\left[A^T PA_d\right]x(k-h)\right\}^{\frac{1}{2}}\Bigg)^{\frac{1}{2}} \\
& -2\left\{x^T(k-h)\left[A_d^T PA\right]x(k)\right\}^{\frac{1}{2}} \left\{x^T(k)\left[A^T PA_d\right]x(k-h)\right\}^{\frac{1}{2}} \\
& -2\left\{x^T(k)\left[A^T PA + Q - P\right]x(k)\right\}^{\frac{1}{2}} \left\{x^T(k-h)\left[A_d^T PA_d - Q\right]x(k-h)\right\}^{\frac{1}{2}} \\
& -2\hat{w}(k)_F \Bigg[\left\{x^T(k)\left[A^T PA_d\right]x(k-h)\right\}^{\frac{1}{2}} + \left\{x^T(k-h)\left[A_d^T PA\right]x(k)\right\}^{\frac{1}{2}} \\
& + \left\{x^T(k-h)\left[A_d^T PA_d - Q\right]x(k-h)\right\}^{\frac{1}{2}} + \left\{x^T(k)\left[A^T PA + Q - P\right]x(k)\right\}^{\frac{1}{2}}\Bigg] \\
& + x^T(k)\left[4A^T PGA + 4A^T G^T PGA\right]x(k) + x^T(k)\left[4A^T PGA_d + 4A^T G^T PGA_d\right]x(k-h) \\
& + x^T(k-h)\left[4A_d^T PGA + 4A_d^T G^T PGA\right]x(k) + x^T(k-h)\left[4A_d^T PGA_d + 4A_d^T G^T PGA_d\right]x(k-h)
\end{aligned} \quad (28)$$

$$g(x(k)) = \begin{bmatrix} \dfrac{1.4x_1^2(k)}{1+x_1^2(k)} \\ \dfrac{x_1(k)}{1+x_1^2(k)+x_2^2(k)} \end{bmatrix}$$

The delay time is assumed to be $h=2$. The initial condition of states x_1 and x_2 are chosen $[0.01 \ \ 0.02]^T$.

The LMIs in H1) and H2) are solved by using Matlab LMI Toolbox and the values of P, Q & G are obtained,

$$P = 10^6 \begin{bmatrix} 0.1492 & 0.6183 \\ 0.6183 & 2.6654 \end{bmatrix}$$

$$Q = \begin{bmatrix} 25.7298 & 59.1751 \\ 59.1751 & 132.4626 \end{bmatrix}$$

$$G = \begin{bmatrix} -0.1707 & 0.0546 \\ 0.0546 & 0.7737 \end{bmatrix},$$

Figures 2 and 3 show the responses of state variables x_1 and x_2. **Figure 4** displays the control input signal. Clearly, simulation results verify our theoretical findings and show the effectiveness of proposed scheme.

6. CONCLUSION

In this paper the problem of stabilization in discrete-time delay nonlinear system is studied, where the system dynamics is assumed to be unknown. The unknown nonlinear functions of the system are approximated through CNN. Two LMIs derived here are sufficient to characterize the controller, which ensures the stability for the resulting closed loop system. A numerical example, with fixed known delay, has been provided to demonstrate the effectiveness and the applicability of the proposed approach.

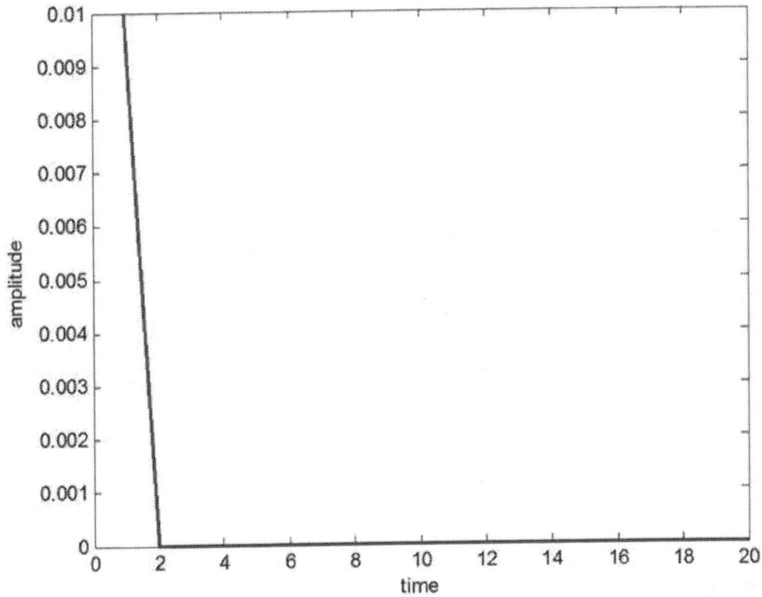

Figure 2. $x_1(k)$ with respect to sample time k (h = 2).

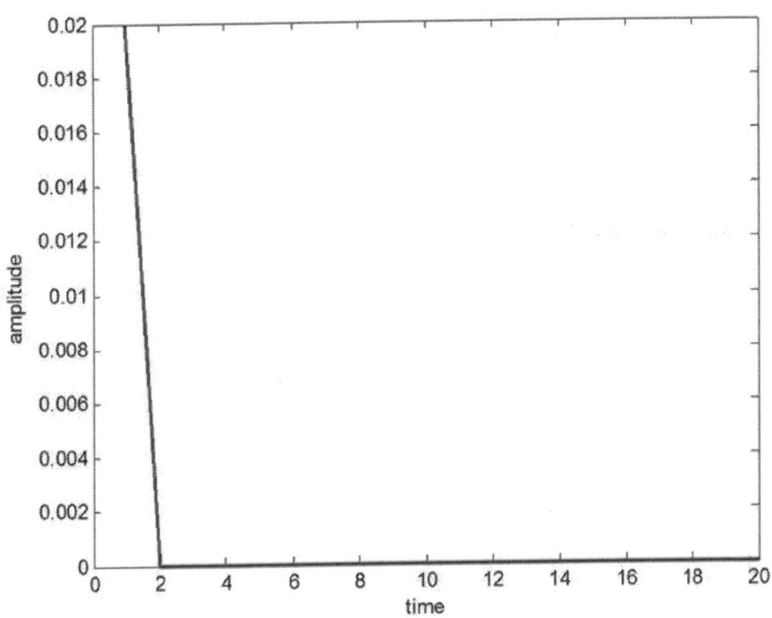

Figure 3. $x_2(k)$ with respect to sample time k (h = 2).

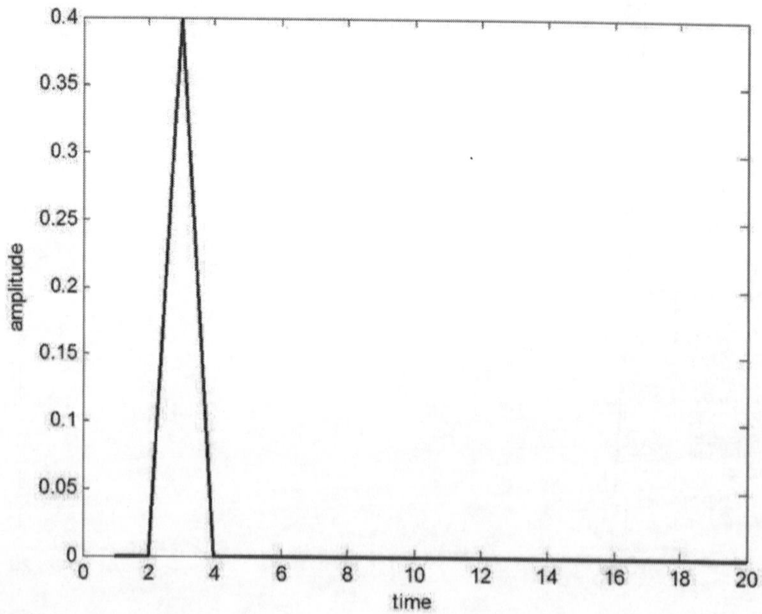

Figure 4. u(k) with respect to sample time k (h = 2).

REFERENCES

1. M. Basin, J. Perez and R. Martinenz-Zuniga, "Optimal Filtering for Nonlinear Polynomial Systems over Linear Observations with Delay," International Journal of Innovative Computing Information and Control, Vol. 2, 2006, pp. 863-874.

2. J. Chen, D. M. Xu and B. Shafai, "On Sufficient Conditions for Stability Independent of Delay," IEEE Transactions on Automatic Control, Vol. 40, No. 9, 1995, pp. 1675-1680.

3. E. Fridman and U. Shaked, "Stability and Guaranteed Cost Control of Uncertain Discrete-Delay Systems," International Journal of Control, Vol. 78, No. 4, 2005, pp. 235-246.

4. C. Lin, Q.-G. Wang and T. H. Lee, "A Less Conservative Robust Stability Test for Linear Uncertain Time-Delay Systems," IEEE Transactions on Automatic Control, Vol. 51, No. 1, 2006, pp. 87-91.

5. Z. Wang, B. Huang and H. Unbehauen, "Robust H_x Observer Design of Linear State Delayed Systems with Parameteric Uncertainty: The Discrete-Time Case," Automatica, Vol. 35, No. 6, 1999, pp. 1161-1167.

6. Y. He, Q. G. Wang, C. Lin and M. Wu, "Delay-RangeDependent Stability for Systems with Time Varying Delay," Automatica, Vol. 43, No. 2, 2007, pp. 371-376.

7. X. Li and C. de Sauza, "Criteria for Robust Stability and Stabilization of Uncertain Linear Systems with State Delay," Automatica, Vol. 33, No. 9, 1997, pp. 1697-1662.

8. V. Kolmanvoskii and J.-P. Richard, "Stability of Some Linear Systems with Delays," IEEE Transactions on Automatic Control, Vol. 44, No. 5, 1999, pp. 984-989.

9. E. Fridman, "New Lyapunov-Krasovskii Functional for Stability of Linear Retarted and Neutral Type Systems," Systems and Control Letters, Vol. 43, No. 4, 2001, pp. 309-319.

10. S.-I. Niculescu, "Delay Effects on Stability: A Robust Control Approach," Lecture Notes in Control and Information Sciences, Vol. 269, Springer-Verlag, London, 2001.

11. E. Fridman and U. Shaked, "An Improved Stabilization Method for Linear Systems with Time-Delay," IEEE Transactions on Automation Control, Vol. 47, No. 11, 2002, pp. 1931-1937.

12. E. Fridman, "Stability of Linear Functional Differential Equations: A New Lyapunov Technique," Proceedings of Mathematical Theory of Networks and Systems, Leuven, 5-9 July 2004.

13. K. J. Astrom and B. Wittenmark, "Computer Controlled Systems: Theory and Design," Prentice-Hall Inc., Englewood Cliffs, 1984.

14. E. Fridman and U. Shaked, "An LMI Approach to Stability of Discrete Delay Systems," Proceedings of 7th IEE European Control Conference, Cambridge, 1-4 September, 2003.

15. Y. S. Lee and W. H. Kwon, "Delay Dependent Robust Stabilization of Uncertain Discrete-Time State-Delayed Systems," 15th Triennial World Congress, Barcelona, 21-26 July 2002.

16. S. S. Ge, F. Hong and T. H. Lee, "Adaptive Neural Network Control of Nonlinear Systems with Unknown Time Delays," IEEE Transactions on Automatic Control, Vol. 48, No. 11, 2003, pp. 2004-2010.

17. S. S. Ge, F. Hong and T. H. Lee, "Adaptive Neural Control of Nonlinear Time-Delay Systems with Unknown Virtual Control Coefficients," IEEE Transactions on Systems, Man, and Cybernetics—Part B: Cybernetics, Vol. 34, No. 1, 2004, pp. 499-516.

18. D. W. C. Ho, J. Li and Y. Niu, "Adaptive Neural Control for a Class of Nonlinearly Parametric Time-Delay Systems," IEEE Transactions on Neural Networks, Vol. 16, No. 3, 2005, pp. 625-635.

19. A. Kulkarni and S. Purwar, "Backstepping Control for a Class of Delayed Nonlinear Systems with Input Constraints," IEEE Conference, Vol. 1, 2008, pp. 274-279.

20. B. Chen, X. P. Liu, K. F. Liu and C. Lin, "Fuzzy-Approximation-Based Adaptive Control of Strict-Feedback Nonlinear Systems with Time Delays," IEEE Transactions on Fuzzy Systems, Vol. 18, No. 5, 2010, pp. 883-891.

21. W. Aggoune, R. Kharel and K. Busawon, "On Feedback Stabilization of Nonlinear Discrete-Time State-Delayed Systems," ECC Conference, Budapest, 23-26 August 2009.

22. H. J. Gao and T. W. Chen, "New Results on Stability of Discrete-Time Systems with Time-Varying State Delay," IEEE Transactions on Automatic Control, Vol. 52, No. 2, 2007, pp. 328-334.

23. S. Ibrir, W. F. Xie and C.-Y. Su, "Observer Design for Discrete-Time Systems Subject to Time Delay Nonlinearities," International Journal of Systems Science, Vol. 37, No. 9, 2006, pp. 629-641.

24. Q.-L. Wei, H.-G. Zhang, D.-R. Liu and Y. Zhao, "An Optimal Control Scheme for a Class of Discrete-Time Nonlinear Systems with t Delays Using Adaptive Dynamic Programming," ACTA Automatica Sinica, Vol. 36, No. 1, 2010, pp. 121-129.

25. K. E. Bouazza, M. Boutayeb and M. Darouach, "State Feedback Stabilization of Discrete-Time Delay Nonlinear Systems," Proceedings of Mathematical Theory of Networks and Systems, Leuven, 5-9 July 2004.

26. E. J. Hartman, J. D. Keelar and J. M. Kowalski, "Layered Neural Networks with Gaussian Hidden Units as Universal Approximation," Neural Computation, Vol. 2, 1990, pp. 210-215.

27. S. Chen, S. A. Billings and P. M. Grant, "Recursive Hybrid Algorithm for Nonlinear System Identification Using Radial Basis Function Networks," International Journal of Control, Vol. 55, No. 5, 1992, pp. 1051-1070.

28. S. V. T. Elanayar and Y. C. Shin, "Radial Basis Function Neural Network for Approximation and Estimation of Nonlinear Stochastic Dynamic Systems," IEEE Transactions on Neural Networks, Vol. 5, No. 4, 1994, pp. 594- 603.

29. J. C. Patra and A. C. Kot, "Nonlinear Dynamic System Identification Using Chebyshev Functional Link Artificial Neural Networks," IEEE Transactions on Systems, Man, and Cybernetics, Vol. 32, No. 4, 2002, pp. 505-511.

30. A. Namatame and N. Ueda, "Pattern Classification with Chebyshev Neural Network," International Journal Neural Network, Vol. 3, 1992, pp. 23-31.

31. T. T. Lee and J. T. Jeng, "The Chebyshev Polynomial Based Unified Model Neural Networks for Functions Approximations," IEEE Transactions on Systems, Man & Cybernetics, Part B, Vol. 28, No. 6, 1998, pp. 925-935.

32. S. Purwar, I. N. Kar and A. N. Jha, "Adaptive Output Feedback Tracking Control of Robot Manipulators Using Position Measurements Only," Expert Systems with Applications, Vol. 34, No. 4, 2008, pp. 2789-2798.

33. S. Purwar, I. N. Kar and A. N. Jha, "On-Line System Identification of Complex Systems Using Chebyshev Neural Networks," Applied Soft Computing, Vol. 7, No. 1, 2005, pp. 364-372.

34. C. Kwan and F. L. Lewis, "Robust Backstepping Control of Nonlinear Systems Using Neural Networks," IEEE Transactions on Systems, Man, and Cybernetics, Vol. 30, No. 6, 2000, pp. 753-766.

4

ADAPTIVE OUTPUT TRACKING FOR NONLINEAR NETWORK CONTROL SYSTEMS WITH TIME-DELAY

Jimin Yu, Haiyan Zeng

College of Automation, Chongqing University of Post and Telecommunications, Chongqing, China

ABSTRACT

The problem of adaptive output tracking is researched for a class of nonlinear network control systems with parameter uncertainties and time-delay. In this paper, a new program is proposed to design a state-feedback controller for this system. For time-delay and parameter uncertainties problems in network control systems, applying the backstepping recursive method, and using Young inequality to process the time-delay term of the systems, a robust adaptive output tracking controller is designed to achieve robust control over a class of nonlinear time-delay network control systems. According to Lyapunov stability theory, Barbalat lemma and Gronwall inequality, it is proved that the designed state feedback controller not only guarantees the state of systems is uniformly bounded, but also ensures the tracking error of the systems converges to a small neighborhood of the origin. Finally, a simulation example for nonlinear network control systems with parameter uncertainties and time-delay is given to illustrate the robust effectiveness of the designed state-feedback controller.

KEYWORDS

Time-Delay; Network Control Systems; Backstepping Design; Adaptive Control; Output Tracking

1. INTRODUCTION

Network control system is a real-time closed-loop feedback control system composed of sensors, controllers, actuators, etc. The advantages of network control systems are its easy installation and maintenance, and its high reliability and flexibility [1,2]. In recent decades, there are lots of progresses in the study of stability of the network control systems [3-6].

However, in the closed-loop control of the network control system, the data transmission process is often produce time-delay. The time-delay of network control system often affects the stability and performance of the system, and may even cause the instability of the entire system [7]. Therefore, the impact of the time-delay on network control system needs to be considered when studying network control systems and designing controllers. In [8], the authors analyzed the source of the time-delay of network control system. For the time-delay problem of network control system, a maximum allowable delay bound satisfying the requirement of stability was proposed in [9], and the maximum delay caused by the network was estimated in [10]. For designing controllers of network control systems, in [11], the authors discussed a class of uncertain systems' adaptive control scheme, and in [12] authors analysis robust stability of networked control systems with uncertainty. Although some progresses are made in linear network control systems, nonlinear network control systems with parameter uncertainties and time-delay needs to be studied. For example, in [13-17], the authors study the problems of adaptive robust control for uncertain systems and highorder uncertain nonlinear systems, and analyze the stability of the systems by Lyapunov stability theory. But these papers did not consider the situation of the systems with time-delay.

Therefore, in this paper, the system is modeled as a class of nonlinear network control system with parameter uncertainties and time-delay. A new program is proposed to design controller for this system, and a robust controller is designed by using the backstepping method. According to Lyapunov stability theory, Barbalat lemma and Gronwall inequality, it is proved that the designed controller not only guarantees the state of nonlinear network control systems with parameter uncertainties and time-delay is uniformly bounded, but also ensures the tracking error of the systems converges to a small neighborhood of the origin. The rest parts of the paper are organized as follows: in Section 2, a class of nonlinear network control system is introduced, and the assumption and lemmas are proposed. In Section 3, the controller is designed by using the backstepping method. In Section 4, a simulation example is presented. Finally, a conclusion is given in Section 5.

2. PROBLEM DESCRIPTION

In this paper, we consider a class of nonlinear network control systems with parameter uncertainties and timedelay, this system is described as

$$\begin{cases} \dot{x}_i = d_i(t,x,u)x_{i+1} + \theta^T \psi_i(\bar{x}_i,t) \\ \quad + g_i(\bar{x}_i,t) + h_i(x_1(t-\tau)) \\ \quad 1 \le i \le n-1 \\ \dot{x}_n = d_n(t,x,u)u + \theta^T \psi_n(\bar{x}_n,t) \\ \quad + g_n(\bar{x}_n,t) + h_n(x_1(t-\tau)) \\ \quad y = x_1 \end{cases} \tag{1}$$

where $\bar{x}_i = [x_1,\cdots,x_i]^T \in R^i (i=1,\cdots,n)$, $u \in R$, and $y \in R$ are respectively the states, the control input and system output, $\theta = [\theta_1,\cdots,\theta_q]^T \in R^q$ is a vector of unknown constant parameters, $d_i(\cdot) \neq 0$, $\psi_i(\cdot)$ and $g_i(\cdot)$ are unknown smooth functions, $h_i(0) = 0$ $(1 \le i \le n)$ is also an unknown smooth functions, τ is time-delay, and $\tau \ge 0$.

The objective of this paper is to design an adaptive feedback controller. The designed controller ensures the state of the closed-loop systems is bounded and the trajectory of output y(t) can asymptotic track reference signal $y_r(t)$.

Assumption 1 For smooth function $d_i(t, x, u)$, $i = 1, \cdots, n$ there exist functions $c_i : R^i - R$ and $\overline{c}_i : R^{i+1} - R$ satisfies $0 < c_i(x_1, \cdots, x_i) \le d_i(t, x, u) \le c_i(x_1, \cdots, x_{i+1})$, $x_{n+1} = u$.

Assumption 2 Because we have $h_i(0) = 0$, then the $h_i(x_1(t))$ can be expressed as $h_i(x_1(t)) = \gamma_i(x_1(t))$, and $\gamma_i(x_1(t))$ satisfies the following assumption

$$\left|\gamma_i(x_1(t))\right| \le \left|p_i(x_1(t))\right|$$

where $p_i(x_1(t))$ is a known and smooth enough function.

Lemma 1 If the real number $a \ge 0$, $b \ge 0$, $m \ge 1$, then there exist the following inequality

$$a \le b + \left(\frac{a}{m}\right)^m \left(\frac{m-1}{b}\right)^{m-1}.$$

Proof for any real number $x \ge 0$, $y > 0$, $n > 0$, by Young inequality, we have

$$xy^n \le \frac{1}{1+n} x^{1+n} + \frac{n}{1+n} y^{1+n}.$$

Let $a = x$, $b = \frac{n}{1+n} y$, $m = n + 1$ then we can release to Lemma 1.

Barbalat lemma [18] If $x(t)$ is a uniformly continuous function, and $\lim_{t \to \infty} \int_0^t x(\tau) d\tau$ exists and is bounded, then $\lim_{t \to \infty} x(\tau) = 0$.

3. ADAPTIVE CONTROLLER DESIGN

In this section, by using the backstepping recursive method, we design a robust adaptive output tracking controller. The designed ideas of this method are described as follows: for the i-th equation of the system, constructed a suitable Lyapunov function, and designed virtual control law α_i, the designed α_i makes the subsystem consist of previous i equations is stable, therefore, in step n, the designed controller u which makes the system consist of n equations stability is the true controller that makes the closed-loop control systems globally stable.

Step 1 Reference signal y_r is a smooth function and bounded, and its derivative $\dot{y}_r$ is also bounded, the output tracking error is defined by $\varepsilon_1 = x_1 - y_r$.

Constructed Lyapunov function as

$V_1 = \frac{1}{2}\varepsilon_1^2 + \frac{1}{2\lambda}\tilde{\theta}^T\tilde{\theta} + \frac{1}{2}\int_{t-\tau}^{t} q(x(s))ds$ where λ are positive, $\tilde{\theta} = \theta - \hat{\theta}$, $\hat{\theta}$ is estimates of the unknown constant parameter θ. Calculating the derivative of V_1 along with system (1), we have

$$\begin{aligned}\dot{V}_1 &= \varepsilon_1\dot{\varepsilon}_1 - \frac{1}{\lambda}\tilde{\theta}^T\dot{\hat{\theta}} + \frac{1}{2}\left[q(x(t)) - q(x(t-\tau))\right] \\ &= \varepsilon_1 d_1 x_2 + \varepsilon_1\left(\hat{\theta}^T\psi_1 + g_1 - \dot{y}_r\right) + \varepsilon_1 h_1\left(x_1(t-\tau)\right) \\ &\quad + \frac{1}{2}\left[q(x(t)) - q(x(t-\tau))\right] - \frac{1}{\lambda}\tilde{\theta}^T\left(\dot{\hat{\theta}} - \lambda\varepsilon_1\psi\right)_1\end{aligned} \tag{2}$$

Because $\dot{y}_r$ is bounded, presence non-negative smooth function $w_1(\varepsilon_1, \hat{\theta})$, satisfies

$$\left|\hat{\theta}^T\psi_1 + g_1 - \dot{y}_r\right| \le w_1\left(\varepsilon_1, \hat{\theta}\right).$$

By Lemma 1, for any real number σ that greater than zero, let $a = |\varepsilon_1| w_1(\varepsilon_1, \hat{\theta})$, b = σ, so that exists a smooth function $\beta_1(\varepsilon_1, \hat{\theta}) \ge 0$, satisfies

$$|\varepsilon_1| w_1\left(\varepsilon_1, \hat{\theta}\right) \le \sigma + \varepsilon_1^2\beta_1\left(\varepsilon_1, \hat{\theta}\right). \tag{3}$$

By using Young inequality, let constant $\xi_1 > 0$ we have

$$\varepsilon_1 h_1\left(x_1(t-\tau)\right) \le \frac{\xi_1^2}{2}\varepsilon_1^2 + \frac{1}{2\xi_1^2}h_1^2\left(x_1(t-\tau)\right), \tag{4}$$

select $q(x(t-\tau)) = \frac{1}{\xi_1^2}h_1^2\left(x_1(t-\tau)\right)$ then we have

$$q(x(t)) = \frac{1}{\xi_1^2} h_1^2(x_1(t)) \le \frac{1}{\xi_1^2} \varepsilon_1^2 \left\| \rho_1(x_1(t)) \right\|^2, \tag{5}$$

where $\rho_1(x_1(t))$ is a smooth function.

Let $z_1 = \lambda\varepsilon_1\psi_1$, Substituting (3), (4), (5) into (2), we have

$$\dot{V}_1 \le \varepsilon_1 d_1 x_2 + \varepsilon_1^2 \beta_1(\varepsilon_1, \hat{\theta}) + \frac{1}{2\xi_1^2} \varepsilon_1^2 \left\| \rho_1(x_1(t)) \right\|^2$$

$$+ \sigma - \frac{1}{\lambda} \tilde{\theta}^T \left(\dot{\hat{\theta}} - z_1 \right) + \frac{\xi_1^2}{2} \varepsilon_1^2$$

Designed virtual controller as

$$\alpha_2 = -\varepsilon_1 \left(\frac{2n + \frac{\xi_1^2}{2} + \beta_1(\varepsilon_1, \hat{\theta})}{c_1(x_1)} \right) = -\varepsilon_1 \phi_1(\cdot)$$

where $f_1(\cdot)$ is smooth function that is greater than zero.

So that we can release to

$$\dot{V}_1 \le -n\varepsilon_1^2 + \varepsilon_1 d_1 x_2 - c_1 \varepsilon_1 \alpha_2 + \sigma - \frac{1}{\lambda} \tilde{\theta}^T \left(\dot{\hat{\theta}} - z_1 \right)$$

$$-\left(n - \frac{1}{2\xi_1^2} \left\| \rho_1(x_1(t)) \right\|^2 \right) \varepsilon_1^2$$

And because $-\varepsilon_1\alpha_2 \ge 0$, by assumption 1, we have

$$\dot{V}_1 \le -n\varepsilon_1^2 - \left(\frac{1}{\lambda} \tilde{\theta} + \eta_1 \right)^T \left(\dot{\hat{\theta}} - z_1 \right)$$

$$-\left(n - \frac{1}{2\xi_1^2} \left\| \rho_1(x_1(t)) \right\|^2 \right) \varepsilon_1^2 + \sigma + \bar{c}_1 |\varepsilon_1| \cdot |x_2 - \alpha_2|$$

where $\eta_1 = 0$.

Step 2 Let $\varepsilon_2 = x_2 - \alpha_2$, constructed Lyapunov function as

$$V_2 = V_1 + \frac{1}{2}\varepsilon_2^{\ 2} + \frac{1}{2}\int_{t-\tau}^{t} q(x(s))ds$$

Calculating the derivative of V_2 along with system (1), we have

$$\begin{aligned}
\dot{V}_2 &= \dot{V}_1 + \varepsilon_2 \dot{\varepsilon}_2 + \frac{1}{2}\left[q(x(t)) - q(x(t-\tau))\right] \\
&\le -n\varepsilon_1^{\ 2} + \sigma - \left(\frac{1}{\lambda}\tilde{\theta} + \eta_1\right)^T \left(\dot{\hat{\theta}} - z_1\right) + \bar{c}_1 |\varepsilon_1| \cdot |x_2 - \alpha_2| \\
&\quad - \left(n - \frac{1}{2\xi_1^{\ 2}}\left\|\rho_1(x_1(t))\right\|^2\right)\varepsilon_1^{\ 2} + \varepsilon_2\left[d_2 x_3 + \theta^T \psi_2 + g_2\right] \\
&\quad + \varepsilon_2\left[-\frac{\partial \alpha_2}{\partial x_1}\left(d_1 x_2 + \theta^T \psi_1 + g_1 + h_1(x_1(t-\tau))\right)\right] \\
&\quad + \varepsilon_2\left[-\frac{\partial \alpha_2}{\partial y_r}\dot{y}_r - \left(\frac{\partial \alpha_2}{\partial \hat{\theta}}\right)^T \dot{\hat{\theta}}\right] + \varepsilon_2 h_2(x_1(t-\tau)) \\
&\quad + \frac{1}{2}\left[q(x(t)) - q(x(t-\tau))\right]
\end{aligned}$$

Let

$$\begin{cases}
\varphi_2 = \psi_2 - \dfrac{\partial \alpha_2}{\partial x_1}\psi_1, \\
z_2 = z_1 + \lambda \varepsilon_2 \varphi_2, \\
\eta_2 = \eta_1 + \varepsilon_2 \dfrac{\partial \alpha_2}{\partial \hat{\theta}},
\end{cases}$$

Then we have

$$\begin{aligned}\dot{V}_2 \leq & -n\varepsilon_1^2 + \bar{c}_1 |\varepsilon_1| \cdot |x_2 - \alpha_2| + \sigma + \varepsilon_2 d_2 x_3 \\ & - \left(n - \frac{1}{2\xi_1^2} \left\| \rho_1 (x_1(t)) \right\|^2 \right) \varepsilon_1^2 + \varepsilon_2 \left(\hat{\theta}^T \varphi_2 + g_2 \right) \\ & + \varepsilon_2 \left[-\frac{\partial \alpha_2}{\partial x_1} \left(d_1 x_2 + g_1 - \left(\frac{\partial \alpha_2}{\partial \hat{\theta}} \right)^T z_2 - \frac{\partial \alpha_2}{\partial y_r} \dot{y}_r \right) \right] \\ & - \left(\frac{1}{\lambda} \tilde{\theta} + \eta_2 \right)^T \left(\dot{\hat{\theta}} - z_2 \right) - \varepsilon_2 \frac{\partial \alpha_2}{\partial x_1} h_1 (x_1(t-\tau)) \\ & + \varepsilon_2 h_2 (x_1(t-\tau)) + \frac{1}{2} \left[q(x(t)) - q(x(t-\tau)) \right]. \end{aligned} \quad (6)$$

There exists a non-negative smooth function $w_2(\varepsilon_1, \varepsilon_2, \hat{\theta})$, satisfies

$$\left| g_2 + \hat{\theta}^T \varphi_2 - \frac{\partial \alpha_2}{\partial x_1} (d_1 x_2 + g_1) - \frac{\partial \alpha_2}{\partial y_r} \dot{y}_r - \left(\frac{\partial \alpha_2}{\partial \hat{\theta}} \right)^T z_2 \right| \leq w_2(\varepsilon_1, \varepsilon_2, \hat{\theta})$$

By Lemma 1, let $a = |\varepsilon_2| w_2(\varepsilon_1, \varepsilon_2, \hat{\theta})$, b = σ so that there exists a smooth function $\beta_2(\varepsilon_1, \varepsilon_2, \hat{\theta}) \geq 0$, satisfies

$$|\varepsilon_2| w_2(\varepsilon_1, \varepsilon_2, \hat{\theta}) \leq \sigma + \varepsilon_2^2 \beta_2(\varepsilon_1, \varepsilon_2, \hat{\theta}). \quad (7)$$

And because $|\varepsilon_1| \cdot |x_2 - \alpha_2| = |\varepsilon_1| \cdot |\varepsilon_2|$, combined with Lemma 1, there exists a smooth function $\tilde{\beta}_2(\varepsilon_1, \varepsilon_2, \hat{\theta}) \geq 0$ satisfies

$$\bar{c}_1(x_1, x_2) |\varepsilon_1| \cdot |x_2 - \alpha_2| \leq \varepsilon_1^2 + \varepsilon_2^2 \tilde{\beta}_2(\varepsilon_1, \varepsilon_2, \hat{\theta}). \quad (8)$$

By using Young inequality, let constant $\xi_2 > 0$, $\mu_2 > 0$, we have

$$\varepsilon_2 h_2(x_1(t-\tau)) \leq \frac{\xi_2^2}{2} \varepsilon_2^2 + \frac{1}{2\xi_2^2} h_2^2(x_1(t-\tau))$$

$$\frac{\partial \alpha_2}{\partial x_1}\varepsilon_2 h_1\left(x_1\left(t-\tau\right)\right) \le \frac{\mu_2^{\ 2}}{2}\varepsilon_2^{\ 2} + \frac{1}{2\mu_2^{\ 2}}h_1^{\ 2}\left(x_1\left(t-\tau\right)\right)$$

Select

$$q\left(x\left(t-\tau\right)\right) = \frac{1}{\xi_2^{\ 2}}h_2^{\ 2}\left(x_1\left(t-\tau\right)\right) - \frac{1}{\mu_2^{\ 2}}h_1^{\ 2}\left(x_1\left(t-\tau\right)\right)$$

then we have

$$\begin{aligned} q\left(x(t)\right) &= \frac{1}{\xi_2^{\ 2}}h_2^{\ 2}\left(x_1\left(t\right)\right) - \frac{1}{\mu_2^{\ 2}}h_1^{\ 2}\left(x_1\left(t\right)\right) \\ &\le \frac{1}{\xi_2^{\ 2}}\varepsilon_1^{\ 2}\left\|\rho_2\left(x_1\left(t\right)\right)\right\|^2 - \frac{1}{\mu_2^{\ 2}}\varepsilon_1^{\ 2}\left\|\rho_1\left(x_1\left(t\right)\right)\right\|^2 \end{aligned}$$

Then we have

$$\begin{aligned} &\varepsilon_2 h_2\left(x_1\left(t-\tau\right)\right) - \varepsilon_2\frac{\partial \alpha_2}{\partial x_1}h_1\left(x_1\left(t-\tau\right)\right) \\ &+\frac{1}{2}\left[q(x(t)) - q\left(x\left(t-\tau\right)\right)\right] \\ \le\ &\frac{\xi_2^{\ 2}}{2}\varepsilon_2^{\ 2} + \frac{1}{2\xi_2^{\ 2}}\varepsilon_1^{\ 2}\left\|\rho_2\left(x_1\left(t\right)\right)\right\|^2 \\ &-\frac{\mu_2^{\ 2}}{2}\varepsilon_2^{\ 2} - \frac{1}{2\mu_2^{\ 2}}\varepsilon_1^{\ 2}\left\|\rho_1\left(x_1\left(t\right)\right)\right\|^2 \\ \le\ &\frac{\xi_2^{\ 2}}{2}\varepsilon_2^{\ 2} + \frac{1}{2\xi_2^{\ 2}}\varepsilon_1^{\ 2}\left\|\rho_2\left(x_1\left(t\right)\right)\right\|^2 \end{aligned} \tag{9}$$

Substituting (7), (8), (9) into (6), we have

$$\begin{aligned} \dot{V}_2 \le &-(n-1)\varepsilon_1^{\ 2} + 2\sigma - \left(\frac{1}{\lambda}\tilde{\theta} + \eta_2\right)^T\left(\dot{\hat{\theta}} - z_2\right) \\ &-\left(n - \frac{1}{2\xi_1^{\ 2}}\left\|\rho_1\left(x_1\left(t\right)\right)\right\|^2 - \frac{1}{2\xi_2^{\ 2}}\left\|\rho_2\left(x_1\left(t\right)\right)\right\|^2\right)\varepsilon_1^{\ 2} \\ &+\varepsilon_2 d_2 x_3 + \frac{\xi_2^{\ 2}}{2}\varepsilon_2^{\ 2} + \varepsilon_2^{\ 2}\left[\beta_2(\cdot) + \tilde{\beta}_2(\cdot)\right] \end{aligned}$$

Designed virtual controller as

$$\alpha_3 = -\varepsilon_2 \left(\frac{n-1+\beta_2(\cdot)+\tilde{\beta}_2(\cdot)+\frac{\xi_2^{\ 2}}{2}}{c_2(x_1,x_2)} \right) = -\varepsilon_2\phi_2(\cdot)$$

where f2(·) is a smooth function that is greater than zero.

By assumption 1, we have

$$\dot{V}_2 \le -(n-1)\left(\varepsilon_1^{\ 2}+\varepsilon_2^{\ 2}\right)-\left(\frac{1}{\lambda}\tilde{\theta}+\eta_2\right)^T\left(\dot{\hat{\theta}}-z_2\right)+2\sigma$$

$$-\left(n-\sum_{j=1}^{2}\frac{1}{2\xi_j^{\ 2}}\left\|\rho_j\left(x_1(t)\right)\right\|^2\right)\varepsilon_1^{\ 2}+\bar{c}_2\left|\varepsilon_2\right|\cdot\left|x_3-\alpha_2\right|$$

Step i After the recursive design step i-1, we can get a group of smooth virtual controller as

$$\begin{aligned}
\alpha_1 &= y_r, & \varepsilon_1 &= x_1-\alpha_1,\\
\alpha_2 &= -\varepsilon_1\phi_1(\cdot), & \varepsilon_2 &= x_2-\alpha_2,\\
&\vdots & &\vdots\\
\alpha_i &= -\varepsilon_{i-1}\phi_{i-1}(\cdot), & \varepsilon_i &= x_i-\alpha_i.
\end{aligned}$$

where smooth function $f_k(\cdot) > 0$, k = 1, ⋯, i − 1.

Constructed Lyapunov function as

$$V_{i-1} = V_{i-2}+\frac{1}{2}\varepsilon_{i-1}^2+\frac{1}{2}\int_{t-\tau}^{t} q\left(x(s)\right)ds$$

The derivative of V_{i-1} as following

$$\dot{V}_{i-1} \leq -(n-i+2)\left(\sum_{j=1}^{i-1}\varepsilon_j^2\right) + \bar{c}_{i-1}|\varepsilon_{i-1}|\cdot|x_i - \alpha_i|$$
$$+(i-1)\sigma - \left(n - \sum_{j=1}^{i-1}\frac{1}{2\xi_j^2}\|\rho_j(x_1(t))\|^2\right)\varepsilon_1^2$$
$$-\left(\frac{1}{\lambda}\tilde{\theta} + \eta_{i-1}\right)^T\left(\dot{\hat{\theta}} - z_{i-1}\right) \tag{10}$$

Similar to step 2, we can prove (10) is also established in the step i.

Constructed Lyapunov function as

$$V_i = V_{i-1} + \frac{1}{2}\varepsilon_i^2 + \frac{1}{2}\int_{t-\tau}^{t} q(x(s))ds$$

Its derivative is given by

$$\dot{V}_i \leq -(n-i+2)\left(\sum_{j=1}^{i-1}\varepsilon_j^2\right) + \bar{c}_{i-1}(\cdot)|\varepsilon_{i-1}|\cdot|x_i - \alpha_i|$$
$$-\left(n - \sum_{j=1}^{i-1}\frac{1}{2\xi_j^2}\|\rho_j(x_1(t))\|^2\right)\varepsilon_1^2$$
$$-\left(\frac{1}{\lambda}\tilde{\theta} + \eta_{i-1}\right)^T\left(\dot{\hat{\theta}} - z_{i-1}\right)$$
$$+\varepsilon_i\left[d_i x_{i+1} + g_i + \theta^T\psi_i - \frac{\partial\alpha_i}{\partial y_r}\dot{y}_r - \left(\frac{\partial\alpha_i}{\partial\hat{\theta}}\right)^T\dot{\hat{\theta}}\right]$$
$$+\varepsilon_i\left[-\sum_{j=1}^{i-1}\left(\frac{\partial\alpha_i}{\partial x_j}\right)\left(d_j x_{j+1} + \theta^T\psi_j + g_j + h_j(x_1(t-\tau))\right)\right]$$
$$+(i-1)\sigma + \varepsilon_i h_i(x_1(t-\tau))$$
$$+\frac{1}{2}\left[q(x(t)) - q(x(t-\tau))\right].$$

Let

$$\begin{cases} \varphi_i = \psi_i - \sum_{j=1}^{i-1} \frac{\partial \alpha_i}{\partial x_j} \psi_j , \\ z_i = z_{i-1} + \lambda \varepsilon_i \varphi_i , \\ \eta_i = \eta_{i-1} + \varepsilon_i \frac{\partial \alpha_i}{\partial \hat{\theta}} , \end{cases}$$

Then we have

$$\begin{aligned} \dot{V}_i \le & -(n-i+2)\left(\sum_{j=1}^{i-1} \varepsilon_j^2 \right) + \bar{c}_{i-1}(\cdot)|\varepsilon_{i-1}| \cdot |x_i - \alpha_i| \\ & - \left(n - \sum_{j=1}^{i-1} \frac{1}{2\xi_j^2} \|\rho_j(x_1(t))\|^2 \right) \varepsilon_1^2 - \left(\frac{1}{\lambda} \tilde{\theta} + \eta_i \right)^T \left(\dot{\hat{\theta}} - z_i \right) \\ & + (i-1)\sigma + \varepsilon_i \left[d_i x_{i+1} - \sum_{j=1}^{i-1} \left(\frac{\partial \alpha_i}{\partial x_j} \right) (d_j x_{j+1} + g_j) \right] \\ & + \varepsilon_i \left[g_i + \hat{\theta}^T \varphi_i - \frac{\partial \alpha_i}{\partial y_r} \dot{y}_r - \lambda \eta_{i-1}{}^T \varphi_i - \left(\frac{\partial \alpha_i}{\partial \hat{\theta}} \right)^T z_i \right] \\ & + \varepsilon_i \left[h_i(x_1(t-\tau)) - \sum_{j=1}^{i-1} \left(\frac{\partial \alpha_i}{\partial x_j} \right) \cdot h_j(x_1(t-\tau)) \right] \\ & + \frac{1}{2} \left[q(x(t)) - q(x(t-\tau)) \right]. \end{aligned} \tag{11}$$

There exists a non-negative smooth function $w_i(\cdot)$, satisfies

$$\left| g_i + \hat{\theta}^T \varphi_i - \sum_{j=1}^{i-1} \left(\frac{\partial \alpha_i}{\partial x_j} \right) (d_j x_{j+1} + g_j) - \lambda \eta_{i-1}{}^T \varphi_i - \frac{\partial \alpha_i}{\partial y_r} \dot{y}_r - \left(\frac{\partial \alpha_i}{\partial \hat{\theta}} \right)^T z_i \right| \le w_i(\cdot)$$

By the Lemma 1, there exists a smooth function $\beta_i(\cdot) \ge 0$ satisfies

$$|\varepsilon_i| w_i(\cdot) \le \sigma + \varepsilon_i^2 \beta_i(\cdot) \tag{12}$$

Similar to step 2, there exists a smooth function $\tilde{\beta}_i(\cdot) \geq 0$ satisfies

$$\bar{c}_{i-1}(\cdot)|\varepsilon_{i-1}|\cdot|x_i-\alpha_i| \leq \sum_{j=1}^{i-1}\varepsilon_j^2+\varepsilon_i^2\tilde{\beta}_i(\cdot) \quad (13)$$

By using Young inequality, let constant $\xi_i > 0$, $\mu_i > 0$, we have

$$\varepsilon_i h_i(x_1(t-\tau)) \leq \frac{\xi_i^2}{2}\varepsilon_i^2+\frac{1}{2\xi_i^2}h_i^2(x_1(t-\tau))$$

$$\varepsilon_i\sum_{j=1}^{i-1}\frac{\partial\alpha_i}{\partial x_j}h_j(x_1(t-\tau)) \leq \frac{(i-1)\mu_i^2}{2}\varepsilon_i^2 + \sum_{j=1}^{i-1}\frac{1}{2\mu_i^2}h_j^2(x_1(t-\tau))$$

select

$$q(x(t-\tau))=\frac{1}{\xi_i^2}h_i^2(x_1(t-\tau))-\sum_{j=1}^{i-1}\frac{1}{\mu_i^2}h_j^2(x_1(t-\tau))$$

then we have

$$q(x(t))=\frac{1}{\xi_i^2}h_i^2(x_1(t))-\sum_{j=1}^{i-1}\frac{1}{\mu_i^2}h_j^2(x_1(t)) \leq \frac{1}{\xi_i^2}\varepsilon_1^2\|\rho_i(x_1(t))\|^2-\sum_{j=1}^{i-1}\frac{1}{\mu_i^2}\varepsilon_1^2\|\rho_j(x_1(t))\|^2$$

Then we have

$$\varepsilon_i h_i\left(x_1(t-\tau)\right)-\varepsilon_i\sum_{j=1}^{i-1}\frac{\partial\alpha_i}{\partial x_j}h_j\left(x_1(t-\tau)\right)$$

$$+\frac{1}{2}\left[q(x(t))-q\left(x(t-\tau)\right)\right]$$

$$\leq\left(\frac{\xi_i^2}{2}-\frac{(i-1)\mu_i^2}{2}\right)\varepsilon_i^2$$

$$+\varepsilon_1^2\left(\frac{1}{2\xi_i^2}\left\|\rho_i\left(x_1(t)\right)\right\|^2-\sum_{j=1}^{i-1}\frac{1}{2\mu_i^2}\left\|\rho_j\left(x_1(t)\right)\right\|^2\right)$$

$$\leq\frac{\xi_i^2}{2}\varepsilon_i^2+\frac{1}{2\xi_i^2}\varepsilon_1^2\left\|\rho_i\left(x_1(t)\right)\right\|^2$$

Then we have

$$\dot{V}_i\leq-(n-i+1)\left(\sum_{j=1}^{i-1}\varepsilon_j^2\right)+\varepsilon_i d_i x_{i+1}+\frac{\xi_i^2}{2}\varepsilon_i^2$$

$$-\left(n-\sum_{j=1}^{i}\frac{1}{2\xi_j^2}\left\|\rho_j\left(x_1(t)\right)\right\|^2\right)\varepsilon_1^2+i\sigma$$

$$-\left(\frac{1}{\lambda}\tilde{\theta}+\eta_i\right)^T\left(\dot{\hat{\theta}}-z_i\right)+\varepsilon_i^2\left[\beta_i(\cdot)+\tilde{\beta}_i(\cdot)\right]$$

Designed virtual controller as

$$\alpha_{i+1}=-\varepsilon_i\left(\frac{n-i+1+\beta_i(\cdot)+\tilde{\beta}_i(\cdot)+\frac{\xi_i^2}{2}}{c_i\left(x_1,\cdots,x_i\right)}\right)=-\varepsilon_i\phi_i(\cdot)$$

where $f_i(\cdot)$ is a smooth function that is greater than zero.

By assumption 1, we have

$$\dot{V}_i\leq-(n-i+1)\sum_{j=1}^{i}\varepsilon_j^2-\left(n-\sum_{j=1}^{i}\frac{1}{2\xi_j^2}\left\|\rho_j\left(x_1(t)\right)\right\|^2\right)\varepsilon_1^2$$

$$+i\sigma+\bar{c}_i\left|\varepsilon_i\right|\cdot\left|x_{i+1}-\alpha_{i+1}\right|-\left(\frac{1}{\lambda}\tilde{\theta}+\eta_i\right)^T\left(\dot{\hat{\theta}}-z_i\right).$$

Step n After repeated recurrence and proof, in the step n, constructed Lyapunov function as

$$V_n = V_{n-1} + \frac{1}{2}\varepsilon_n^{\ 2} + \frac{1}{2}\int_{t-\tau}^{t} q(x(s))ds$$

Its derivative is given by

$$\begin{aligned}\dot{V}_n \le & -\sum_{j=1}^{n}\varepsilon_j^{\ 2} + n\sigma - \left(\frac{1}{\lambda}\tilde{\theta} + \eta_n\right)^T \left(\dot{\hat{\theta}} - z_n\right) \\ & -\left(n - \sum_{j=1}^{n}\frac{1}{2\xi_j^{\ 2}}\left\|\rho_j\left(x_1(t)\right)\right\|^2\right)\varepsilon_1^{\ 2} \\ & +\bar{c}_n\left|\varepsilon_n\right|\cdot\left|x_{n+1} - \alpha_{n+1}\right| \end{aligned} \tag{14}$$

From (14), we can obtain adaptive control law u and parameter $\dot{\hat{\theta}}$ following as

$$\begin{aligned} u &= \alpha_{n+1} \\ &= -\varepsilon_n \left(\frac{1 + \beta_n(\cdot) + \tilde{\beta}_n(\cdot) + \frac{\xi_n^{\ 2}}{2}}{c_n(x_1, \cdots, x_n)} \right) = -\varepsilon_n \phi_n(\cdot) \end{aligned} \tag{15}$$

$$\dot{\hat{\theta}} = z_n \tag{16}$$

where $f_n(\cdot)$ is a smooth function that is greater than zero.

Then, we have

$$\dot{V}_n \le -\sum_{j=1}^{n}\varepsilon_j^{\ 2} + n\sigma - \left(n - \sum_{j=1}^{n}\frac{1}{2\xi_j^{\ 2}}\left\|\rho_j\left(x_1(t)\right)\right\|^2\right)\varepsilon_1^{\ 2}.$$

When n is large enough, then we have

$$\dot{V}_n \le -\sum_{j=1}^{n} \varepsilon_j^2 + n\sigma$$

Select $p_n = \sum_{j=1}^{n} \varepsilon_j^2$ then we have $V_n(t) - V_0(t) \le -\int_0^t (p_n)\,dt + n\sigma t$.

Therefore, we get $0 \le \int_0^t (p_n)\,dt \le V_0(t) + n\sigma t < \infty$.

By Barbalat lemma, we get $\lim_{t\to\infty} p_n = 0$, and then we have $\lim_{t\to\infty} \varepsilon_j = 0$, j = 1, ···, n. So that we get $\lim_{t\to\infty} |y(t) - y_r(t)| = 0$.

So that the entire design procedure is reasonable.

Theorem 1 Considering closed-loop systems (1), under assumption and Lemma, there exist a state feedback control law u and control law parameter $\dot{\hat{\theta}}$. The closedloop system is bounded for all allowable uncertainties and the output tracking error converges to a relatively small area, which satisfies

$$|y(t) - y_r(t)| = |\varepsilon_1| \le \sqrt{2A/B} + \sqrt{2V_n(0)e^{-Bt}}.$$

Proof

$$\dot{V}_n \le -\sum_{j=1}^{n} \varepsilon_j^2 + n\sigma \le -BV_n + A$$

where

$$B = \min\{2, \lambda\sigma\} > 0$$

$$A = n\sigma + \frac{n}{2}\int_{t-d}^{t} q(x(s))ds.$$

By Gronwall inequality, we have

$$\begin{aligned} V_n(t) &\le A/B + [V_n(0) - A/B]e^{-Bt} \\ &\le A/B + V_n(0)e^{-Bt}. \end{aligned}$$

And because

$$\frac{1}{2}\varepsilon_j^2 \le V_n(t) \le A/B + V_n(0)e^{-Bt}$$ So that we have

$$|y(t) - y_r(t)| = |\varepsilon_1| \le \sqrt{2A/B} + \sqrt{2V_n(0)e^{-Bt}}$$.

In summary, for any real number $\varepsilon_0 > 0$, in limited time $T > 0$, the closed-loop system satisfies

$$|y(t) - y_r(t)| < \varepsilon_0, \forall t \ge T > 0$$.

4. SIMULATION EXAMPLE

In order to show the effectiveness of the design scheme, we choose the nonlinear network control system with parameter uncertainties and time-delay as following:

$$\begin{cases} \dot{x}_1 = x_2 + \theta^T x_1 + x_1(t-\tau) \\ \dot{x}_2 = u + x_1(t-\tau) \\ y = x_1 \end{cases} \tag{17}$$

In the simulation, for the closed-loop system (17), we choose the reference signal $y_r(t) = \sin t$, time-delay $\tau = 0.01s$, $\theta = 0.2$, $\xi_1 = 1$, $\xi_2 = 2$, $\sigma = 0.02$, $\lambda = 1$, the initial conditions $x_1(0) = 1$, $x_2(0) = 0.5$, $\hat{\theta}(0) = 0.1$, According to (15) and (16), the control law u and the parameter of control law $\dot{\hat{\theta}}$ following as

$$u = -(x_2 - \alpha_2)\cdot\left(1 + \frac{w_2^2}{4\sigma} + \frac{1}{4} + \frac{\xi_2^2}{2}\right)$$

$$= \left[(y_r - x_1)\left(4 + \frac{\xi_1^2}{2} + \frac{\left(\hat{\theta}^T x_1 - \dot{y}_r\right)^2}{4\sigma}\right) - x_2\right]$$

$$\cdot\left[1 + \frac{\left(\left(\frac{\partial\alpha_2}{\partial x_1}\left(\hat{\theta}^T x_1 + x_2\right)\right) + \left(\frac{\partial\alpha_2}{\partial y_r}\dot{y}_r\right) + \left(\frac{\partial\alpha_2}{\partial\hat{\theta}}\right)^T\dot{\hat{\theta}}\right)^2}{4\sigma}\right.$$

$$\left. + \frac{1}{4} + \frac{\xi_2^2}{2}\right],$$

$$\dot{\hat{\theta}} = \lambda\left[x_1(x_1 - y_r) - \varepsilon_2\left(\frac{\partial\alpha_2}{\partial x_1}\right)x_1\right]$$

$$= \lambda\left[x_1(x_1 - y_r)\right.$$

$$\left. - \left(x_2 - (y_r - x_1)\left(4 + \frac{\xi_1^2}{2} + \frac{(\hat{\theta}^T x_1 - \dot{y}_r)^2}{4\sigma}\right)\right)\cdot\left(\frac{\partial\alpha_2}{\partial x_1}\right)x_1\right]$$

The simulation results are shown as in Figures 1 and 2. It can be observed that the output of closed-loop system can track the reference signal well, and the tracking error converges to a small neighborhood of the origin. Therefore the robust adaptive controller is effective.

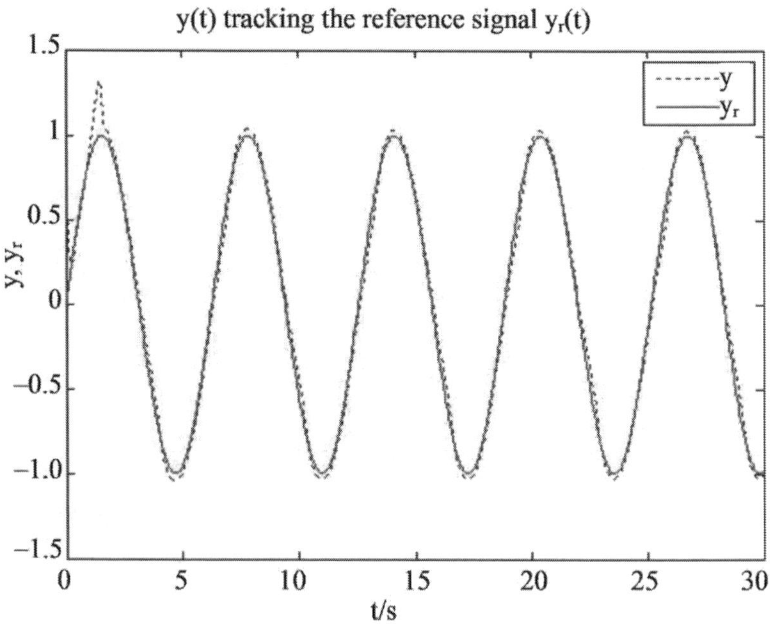

Figure 1. Output y(t) and reference signal $y_r(t)$.

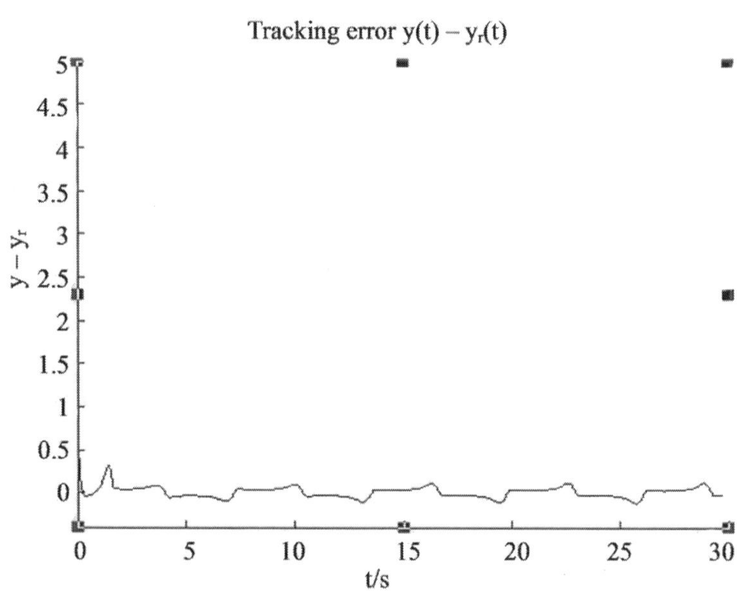

Figure 2. Output tracking error y(t) – $y_r(t)$.

5. CONCLUSION

By using the backstepping method, we design a controller for nonlinear network system with parameter uncertainties and time-delay. Through theoretical analysis, it is shown that the designed robust adaptive output tracking controller is feasible. The simulation results further expressed the effectiveness of the scheme.

ACKNOWLEDGEMENTS

This work was supported in part by the Natural Science Foundation of Chongqing (CSTC) under Grant No. 2009BB3280, and the National Natural Science Foundation of China under Grant No. 60873200.

REFERENCES

1. J. P. Hespanha, P. Naghahtabrizi and Y. G. Xu, "A Survey of Recent Results in Networked Control Systems," Proceedings of the IEEE, Vol. 95, No. 1, 2007, pp. 138- 162.

2. L. A. Montestruque and E. J. Antsaklis, "On the ModelBased Control of Networked Systems," Automatica, Vol. 39, No. 10, 2003, pp. 1837-1843.

3. G. C. Walsh, Y. Hong and L. G. Bushnell, "Stability Analysis of Networked Control Systems," IEEE Transactions on Control Systems Technology, Vol. 10, No. 3, 2002, pp. 438-446.

4. A. V. Savkin, "Analysis and Synthesis of Networked Control Systems: Topologicall Entropy, Observability, Robustness and Optimal Control," Automatica, Vol. 41, No. 1, 2006, pp. 51-62.

5. D. Yue, Q. L. Han and C. Peng, "State Feedback Controller Design of Networked Control Systems," IEEE Transactions on Circuits and Systems II: Express Briefs, Vol. 51, No. 11, 2004, pp. 640-644.

6. P. V. Zhivoglyadov and R. H. Middleton, "Networked Control Design for Linear Systems," Automatica, Vol. 39, No. 4, 2003, pp. 743-750

7. T. Yodyium and M. Y. Chow, "Control Methodologies in Networked Control Systems," Control Engineering Practice, Vol. 11, No. 10, 2003, pp. 1099-1111.

8. W. Zhang, M. S. Branicky and S. M. Phillips, "Stability of Networked Control Systems," IEEE Control Systems Magazine, Vol. 21, No. 1, 2001, pp. 84-99.

9. H. S. Park, Y. H. Kim, D. S. Kim and W. H. Kwon, "A Scheduling Method for Network Based Control Systems," IEEE Transactions on Control Systems Technology, Vol. 10, No. 3, 2002, pp. 318-330.

10. D. S. Kim, Y. S. Lee, W. H. Kwon and H. S. Park, "Maximum Allowable Delay Bounds of Networked Control Systems," Control Engineering Practice, Vol. 11, No. 11, 2003, pp. 1301-1313.

11. D. Liberzon and J. P. Hespanha, "Stabilization of Nonlinear Systems with Limited Information Feedback," IEEE Transaction on Automatic Control, Vol. 50, No. 6, 2005, pp. 910-915.

12. D. Yue, Q. Han and J. Lam, "Network-Based Robust H_∞ Control of Systems with Uncertainty," Automatica, Vol. 41, No. 6, 2005, pp. 999-1007.

13. L. Wei and R. Pongvuthithum, "Nonsmooth Adaptive Stabilization of Cascaded Systems with Nonlinear Parameterization via Partial-State Feedback," IEEE Transaction on Automatic Control, Vol. 48, No. 10, 2003, pp. 1809- 1816.

14. Z. Y. Sun and Y. G. Liu, "State-Feedback Adaptive Stabilizing Control Design for a Class of High-Order Nonlinear Systems with Unknown Control Coefficients," Journal of Systems Science and Complexity, Vol. 20, No. 10, 2007, pp. 350-361.

15. Z. Y. Sun and Y. G. Liu, "Adaptive Stabilization for a Large Class of High-Order Uncertain Nonlinear Systems," International Journal of Control, Vol. 82, No. 7, 2009, pp. 1275-1287.

16. Z. Y. Sun, Y. G. Liu and Z. G. Liu, "An Adaptive Controller for a Class of High-Order Nonlinear Systems with Unknown Control Coefficients," Proceedings of the 30th Chinese Control Conference, Yantai, 22-24 July 2011, pp. 266-271.

17. N. B. He, C. S. Jiang and Q. Gao, "Adaptive Backstepping Control for a Class of Nonlinear Systems," 3rd International Conference on Measuring Technology and Mechatronics Automation, Shanghai, 6-7 January 2011, pp. 322-325.

18. N. B. He, C. S. Jiang and Q. Gao, "Robust Adaptive Backstepping Control of Nonlinear Systems with Uncertainty," Journal of Applied Sciences Electronics and Information Engineering, Vol. 26, No. 6, 2008, pp. 650- 654.

5

TIME-DELAY ESTIMATION IN DISPERSED SPECTRUM COGNITIVE RADIO SYSTEMS

Fatih Kocak, Hasari Celebi, Sinan Gezici, Khalid A. Qaraqe, Huseyin Arslan and H. Vincent Poor

1 Department of Electrical and Electronics Engineering, Bilkent University, Bilkent, 06800 Ankara, Turkey

2 Department of Electrical and Computer Engineering, Texas A&M University at Qatar, 23874 Doha, Qatar

3 Department of Electrical Engineering, University of South Florida, 4202 E Fowler Avenue, Tampa, FL 33620, USA

4 Department of Electrical Engineering, Princeton University, Princeton, NJ 08544, USA

ABSTRACT

Time-delay estimation is studied for cognitive radio systems, which facilitate opportunistic use of spectral resources. A two-step approach is proposed to obtain accurate time-delay estimates of signals that occupy multiple dispersed bands simultaneously, with significantly lower computational complexity than the optimal maximum likelihood (ML) estimator. In the first step of the proposed approach, an ML estimator is used for each band of the signal in order to estimate the unknown parameters of the signal occupying that band. Then, in the second step, the estimates from the first step are combined in various ways in order to obtain the final time-delay estimate. The combining techniques that are used in the second step are called optimal combining, signal-to-noise ratio (SNR) combining, selection combining, and equal combining.

It is shown that the performance of the optimal combining technique gets very close to the Cramer-Rao lower bound at high SNRs. These combining techniques provide various mechanisms for diversity combining for time-delay estimation and extend the concept of diversity in communications systems to the time-delay estimation problem in cognitive radio systems. Simulation results are presented to evaluate the performance of the proposed estimators and to verify the theoretical analysis.

1. INTRODUCTION

Cognitive radio is a promising approach to implement intelligent wireless communications systems [1–8]. Cognitive radios can be perceived as more capable versions of software defined radios in the sense that they have sensing, awareness, learning, adaptation, goal driven autonomous operation, and reconfigurability features [9, 10]. Thanks to these features, radio resources, such as power and bandwidth, can be used more efficiently [1]. Especially since the electromagnetic spectrum is a precious resource, it must not be wasted. The recent spectrum measurement campaigns in the United States [11] and Europe [12] show that the spectrum is under-utilized; hence, opportunistic use of unoccupied frequency bands is highly desirable.

Cognitive radio provides a solution to the problem of inefficient spectrum utilization by using the vacant frequency spectrum over time in a certain geographical region. In other words, a cognitive radio system can opportunistically use the available spectrum of a legacy system without interfering with the licensed users of that spectrum [2, 3]. In order to facilitate such opportunistic spectrum utilization, it is important that cognitive radio devices are aware of their positions, and monitor the environment continuously. These location and environmental awareness features of cognitive radios have been studied extensively in the literature [10, 13–19]. In [13], the concept of cognitive radar is introduced, which provides information related to the objects in an environment; that is, it performs environmental sensing. In [14], a radio environment mapping method for cognitive radio networks is studied. Conceptual models for location and environmental awareness engines and cycles are proposed in [10, 15, 16] for cognitive radio systems. Also, [18] introduces the concept of a topology engine for cognitive radios by studying topology information characterization and its applications to cognitive radio networks.

The location awareness feature of cognitive radios can be used in many network optimization applications, such as location-assisted spectrum management, network planning, handover, routing, dynamic channel allocation, and power control [8, 20]. Location awareness requires that a cognitive radio device performs accurate estimation of its position. One possible way of obtaining position information is to use the Global Positioning System (GPS) technology in cognitive radio systems. However, this is not a very efficient or cost-effective solution [17]. As another approach, cognitive radio devices can estimate position-related parameters of signals traveling between them in order to estimate their positions [17, 21]. Among various position related parameters, the time-delay parameter commonly provides accurate position information with reasonable complexity [21, 22]. The main focus of this study is time-delay estimation in cognitive radio systems. In other words, the aim is to propose techniques for accurate time-delay estimation in dispersed spectrum systems in order to provide accurate location information to cognitive users. Since the accuracy of location estimation increases as the accuracy of time-delay estimation increases, design of time-delay estimators with high accuracy and reasonable complexity is crucial for the location awareness feature of a cognitive radio system [21].

Time-delay estimation in cognitive radio systems differs from conventional time-delay estimation mainly due to the fact that a cognitive radio system can transmit and receive over multiple dispersed bands. In other words, since a cognitive radio device can utilize the spectral holes of a legacy system, it can have a spectrum that consists of multiple bands that are dispersed over a wide range of frequencies (cf. Figure 1). In [23], the theoretical limits on time-delay estimation are studied for dispersed spectrum cognitive radio systems, and the effects of carrier frequency offset (CFO) and modulation schemes of training signals on the accuracy of time-delay estimation are quantified. The expressions for the theoretical limits indicate that frequency diversity can be utilized in time-delay estimation. Similarly, the effects of spatial diversity on time-delay estimation are studied in [24] for single-input multiple-output (SIMO) systems. In addition, the effects of multiple antennas on time-delay estimation and synchronization problems are investigated in [25].

In this paper, time-delay estimation is studied for dispersed spectrum cognitive radio systems. First, it is observed that maximum likelihood (ML) estimation is not very practical for time-delay estimation in such systems. Then, a two-step time-delay estimation approach is proposed in order to provide accurate time-delay estimation with significantly lower computational complexity than that of the optimal ML estimator.

In the proposed scheme, the receiver consists of multiple branches and each branch processes the part of the received signal that occupies the corresponding frequency band. An ML estimator is used in each branch in order to estimate the unknown parameters of the signal observed in that branch. Then, in the second step, the estimates from all the branches are combined to obtain the final time-delay estimate. Various techniques are proposed for the combining operation in the second step: Optimal combining, signal-to-noise ratio (SNR) combining, selection combining, and equal combining. The biases and variances of the time-delay estimators that employ these combining techniques are investigated. It is shown that the optimal combining technique results in a mean-squared error (MSE) that approximates the Cramer-Rao lower bound (CRLB) at high SNRs. Simulation results are provided in order to compare the performance of the proposed time-delay estimators. In a more generic perspective, this study focuses on the utilization of frequency diversity for a parameter estimation problem. Therefore, the proposed estimators can be applied to other systems that have frequency diversity as well.

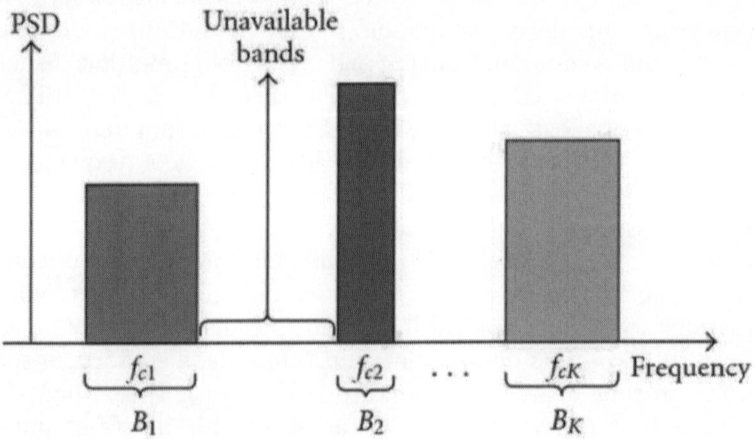

Figure 1 Illustration of dispersed spectrum utilization in cognitive radio systems.

The remainder of the paper is organized as follows. In Section 2, the signal model is introduced and the signal at each branch of the receiver is described. In Section 3, the optimal ML receiver is obtained, and the CRLBs on time-delay estimation in dispersed spectrum cognitive radio systems are described. The proposed two-step time-delay estimation approach is studied in Section 4. Then, in Section 5, the optimality properties of the proposed time-delay estimators are investigated. Finally, simulation results are presented in Section 6, and concluding remarks are made in Section 7.

2. SIGNAL MODEL

A cognitive radio system that occupies K dispersed frequency bands is considered as shown in Figure 1. The transmitter sends a signal occupying all the K bands simultaneously, and the receiver aims to calculate the time-delay of the incoming signal.

One approach for designing such a system involves the use of orthogonal frequency division multiplexing (OFDM). In this approach, the received signal is considered as a single OFDM signal with zero coefficients at the subcarriers corresponding to the unavailable bands [26–28]. Then, the signal can be processed as in conventional OFDM receivers. The main drawback of this approach is that it requires processing of very large bandwidths when the available spectrum is dispersed over a wide range of frequencies. Therefore, the design of RF components, such as filters and low-noise amplifiers (LNAs), can become very complex and costly, and result in components with high power consumption [29]. In such scenarios, it can be more practical to process the received signal in multiple branches, as shown in Figure 2. In that case, each branch processes one available band and down-converts the signal according to the center frequency of that band. Therefore, signals with narrower bandwidths can be processed at each branch.

For the receiver model in Figure 2, the baseband representation of the received signal in the i th branch can be modeled as

$$r_i(t) = \alpha_i e^{j\omega_i t} s_i(t-\tau) + n_i(t), \tag{1}$$

for $i = 1, \ldots, K$, where τ is the time-delay of the signal, $\alpha_i = a_i e^{j\phi_i}$ and ω_i represent, respectively, the channel coefficient and the CFO for the signal in the i th branch, $s_i(t)$ is the baseband representation of the transmitted signal in the i th band, and $n_i(t)$ is modeled as complex white Gaussian noise with independent components, each having spectral density σ_i^2.

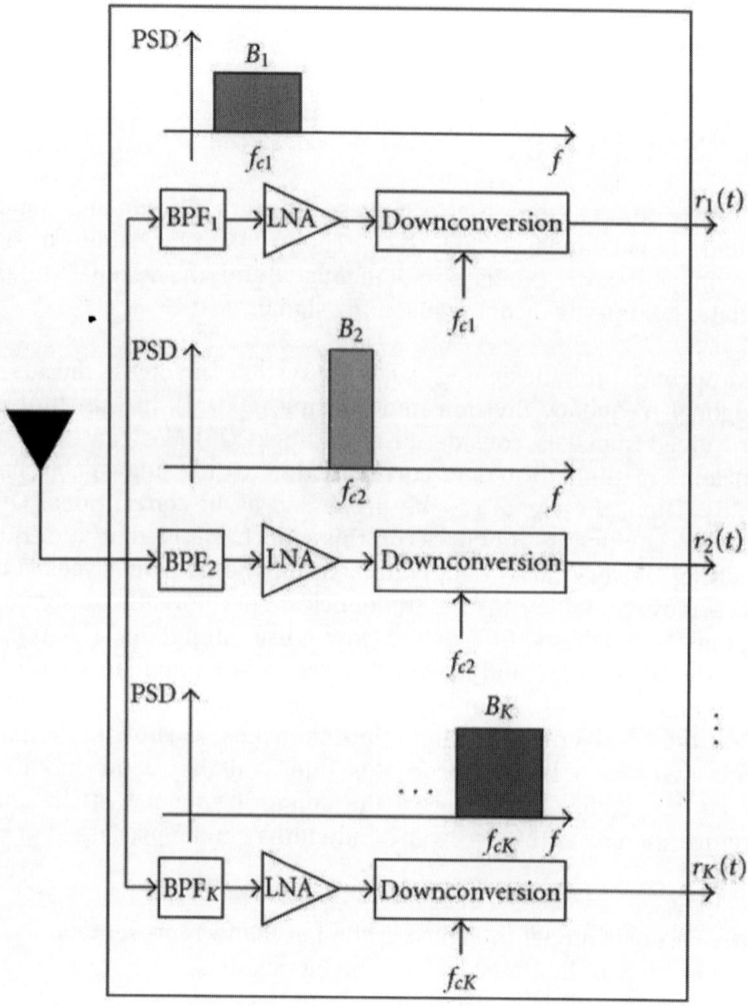

Figure 2 Block diagram of the front-end of a cognitive radio receiver, where BPF and LNA refer to band-pass filter and low-noise amplifier, respectively.

The signal model in (1) assumes that the signal in each branch can be modeled as a narrowband signal. Hence, a single complex channel coefficient is used to represent the fading of each signal.

The system model considered in this study falls within the framework of cognitive radio systems, since the cognitive user first needs to detect the available frequency bands, and then to adapt its receiver parameters accordingly. Therefore, the spectrum sensing and adaptation features of cognitive systems are assumed for the considered system in this study [9, 10].

3. OPTIMAL TIME-DELAY ESTIMATION AND THEORETICAL LIMITS

Accurate estimation of the time-delay parameter τ in (1) is quite challenging due to the presence of unknown channel coefficients and CFOs. For a system with K bands, there are $3K$ nuisance parameters. In other words, the vector $\boldsymbol{\theta}$ of unknown parameters can be expressed as

$$\boldsymbol{\theta} = \left[\tau \;\; a_1 \cdots a_K \;\; \phi_1 \cdots \phi_K \;\; \omega_1 \cdots \omega_K\right]. \tag{2}$$

When the signals in (1) are observed over the interval $[0, T]$, the log-likelihood function for $\boldsymbol{\theta}$ is given by [30]

$$\Lambda(\boldsymbol{\theta}) = c - \sum_{i=1}^{K} \frac{1}{2\sigma_i^2} \int_0^T \left| r_i(t) - \alpha_i e^{j\omega_i t} s_i(t-\tau) \right|^2 dt, \tag{3}$$

where c is a constant that is independent of $\boldsymbol{\theta}$ (the unknown parameters are assumed to be constant during the observation interval). Then, the ML estimate for $\boldsymbol{\theta}$ can be obtained from (3) as [23]

$$\hat{\boldsymbol{\theta}}_{\mathrm{ML}} = \arg\max_{\boldsymbol{\theta}} \left\{ \sum_{i=1}^{K} \frac{1}{\sigma_i^2} \int_0^T \mathcal{R}\left\{ \alpha_i^* e^{-j\omega_i t} r_i(t) s_i^*(t-\tau) \right\} dt - \sum_{i=1}^{K} \frac{E_i |\alpha_i|^2}{2\sigma_i^2} \right\}, \tag{4}$$

where $E_i = \int_0^T | s_i(t-\tau) |^2 dt$ is the signal energy, and $\mathcal{R}$ represents the operator that selects the real part of its argument.

It is observed from (4) that the ML estimator requires an optimization over a $(3K+1)$-dimensional space, which is quite challenging in general. Therefore, the aim of this study is to propose low-complexity time-delay estimation algorithms with comparable performance to that of the ML estimator in (4). In other words, accurate time-delay estimation algorithms are studied under practical constraints on the processing power of the receiver. Since the ML estimator is difficult to implement, the performance comparisons will be made with respect to the theoretical limits on time-delay estimation (an ML estimator achieves the CRLB asymptotically under certain conditions [30]). In [23], the CRLBs on the MSEs of unbiased time-delay estimators are obtained for the signal model in (1). When the baseband representation of the signals in different branches is of the form $s_i(t) = \Sigma_l d_{i,l} p_i(t - lT_i)$, where $d_{i,l}$ denotes the complex training data and $p_i(t)$ is a pulse with duration T_i, the CRLB is expressed as

$$\mathrm{E}\left\{(\hat{\tau}-\tau)^2\right\} \geq \left(\sum_{i=1}^{K} \frac{a_i^2}{\sigma_i^2}\left(\tilde{E}_i - \frac{\left(\hat{E}_i^{\mathrm{R}}\right)^2}{E_i}\right)\right)^{-1}, \tag{5}$$

where

$$\tilde{E}_i = \int_0^T | s_i'(t-\tau) |^2 dt, \tag{6}$$

$$\hat{E}_i^{\mathrm{R}} = \int_0^T \mathcal{R}\{s_i'(t-\tau) s_i^*(t-\tau)\} dt, \tag{7}$$

with $s'(t)$ representing the first derivative of $s(t)$. In the special case of $|d_{i,l}| = |d_i|$ for all l and $p_i(t)$ satisfying $p_i(0) = p_i(T_i)$ for $i = 1, \ldots, K$, (5) becomes [23]

$$\mathrm{E}\left\{(\hat{\tau}-\tau)^2\right\} \geq \left(\sum_{i=1}^{K} \frac{\tilde{E}_i a_i^2}{\sigma_i^2}\right)^{-1}. \tag{8}$$

It is observed from (5) and (8) that frequency diversity can be useful in time-delay estimation. For example, when one of the bands is in a deep fade (i.e., small a_i^2), some other bands can still be in good condition to facilitate accurate time-delay estimation.

4. TWO-STEP TIME-DELAY ESTIMATION AND DIVERSITY COMBINING

Due to the complexity of the ML estimator in (4), a two-step time-delay estimation approach is proposed in this paper, as shown in Figure 3. Two-step approaches are commonly used in optimization/estimation problems in order to provide suboptimal solutions with reduced computational complexity [31, 32]. In the proposed estimator, each branch of the receiver performs estimation of the time-delay, the channel coefficient, and the CFO related to the signal in that branch. Then, the estimates from all the branches are used to obtain the final time-delay estimate as shown in Figure 3. In the following sections, the details of the proposed approach are explained, and the utilization of frequency diversity in time-delay estimation is explained.

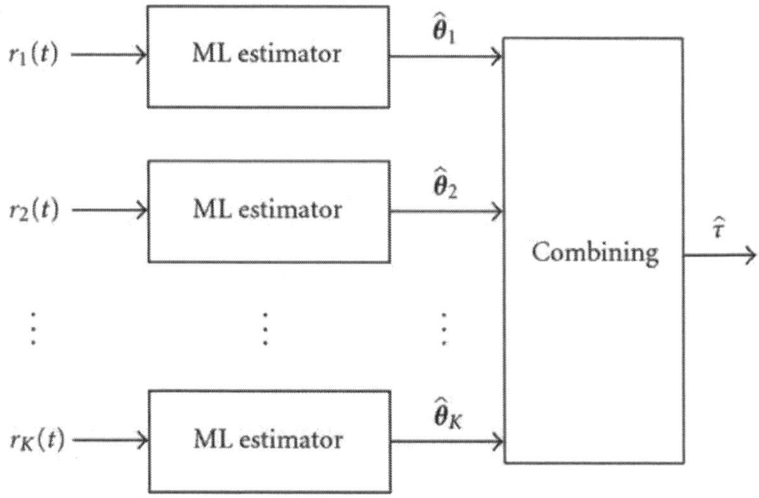

Figure 3 The block diagram of the proposed time-delay estimation approach. The signals $r_1(t),\ldots,r_K(t)$ are obtained at the front-end of the receiver as shown in Figure 2.

4.1. First Step: Parameter Estimation at Different Branches

In the first step of the proposed approach, the unknown parameters of each received signal are estimated at the corresponding receiver branch according to the ML criterion (cf. Figure 3). Based on the signal model in (1), the likelihood function at branch i can be expressed as

$$\Lambda_i(\boldsymbol{\theta}_i) = c_i - \frac{1}{2\sigma_i^2}\int_0^T \left| r_i(t) - \alpha_i e^{j\omega_i t} s_i(t-\tau) \right|^2 dt, \tag{9}$$

for $i = 1, \ldots, K$, where $\boldsymbol{\theta}_i = [\tau\ a_i\ \phi_i\ \omega_i]$ represents the vector of unknown parameters related to the signal at the i th branch, $r_i(t)$, and c_i is a constant that is independent of $\boldsymbol{\theta}_i$.

From (9), the ML estimator at branch i can be stated as

$$\hat{\boldsymbol{\theta}}_i = \arg\min_{\theta_i} \int_0^T \left| r_i(t) - \alpha_i e^{j\omega_i t} s_i(t-\tau) \right|^2 dt, \tag{10}$$

where $\hat{\boldsymbol{\theta}}_i = [\hat{\tau}_i\ \hat{a}_i\ \hat{\phi}_i\ \hat{\omega}_i]$ is the vector of estimates at the i th branch. After some manipulation, the solution of (10) can be obtained as

$$\left[\hat{\tau}_i\ \hat{\phi}_i\ \hat{\omega}_i\right] = \arg\max_{\phi_i,\omega_i,\tau_i} \left| \int_0^T \mathcal{R}\left\{ r_i(t) e^{-j(\omega_i t+\phi_i)} s_i^*(t-\tau_i) \right\} dt \right|, \tag{11}$$

$$\hat{a}_i = \frac{1}{E_i}\int_0^T \mathcal{R}\left\{ r_i(t) e^{-j(\hat{\omega}_i t+\hat{\phi}_i)} s_i^*(t-\hat{\tau}_i) \right\} dt. \tag{12}$$

In other words, at each branch, optimization over a three-dimensional space is required to obtain the unknown parameters. Compared to the ML estimator in Section 3, the optimization problem in (4) over $(3K+1)$ variables is reduced to K optimization problems over three variables, which results in a significant amount of reduction in the computational complexity.

In the absence of CFO; that is, $\omega_i = 0$ for all i , (11) and (12) reduce to

$$\left[\hat{\tau}_i\ \hat{\phi}_i\right] = \arg\max_{\phi_i,\tau_i}\left|\int_0^T \mathcal{R}\left\{r_i(t)e^{-j\phi_i}s_i^*(t-\tau_i)\right\}dt\right|, \tag{13}$$

$$\hat{a}_i = \frac{1}{E_i}\int_0^T \mathcal{R}\left\{r_i(t)e^{-j\hat{\phi}_i}s_i^*(t-\hat{\tau}_i)\right\}dt. \tag{14}$$

In that case, the optimization problem at each branch is performed over only two dimensions. This scenario is valid when the carrier frequency of each band is known accurately.

4.2. Second Step: Combining Estimates from Different Branches

After obtaining K different time-delay estimates, $\hat{\tau}_1,\ldots,\hat{\tau}_K$, in (11), the second step combines those estimates according to one of the criteria below and makes the final time-delay estimate (cf. Figure 3).

4.2.1. Optimal Combining

According to the "optimal" combining criterion (the optimality properties of this combining technique are investigated in Section 5), the time-delay estimate is obtained as

$$\hat{\tau} = \frac{\sum_{i=1}^K \kappa_i\hat{\tau}_i}{\sum_{i=1}^K \kappa_i}, \tag{15}$$

where $\hat{\tau}_i$ is the time-delay estimate of the i th branch, which is obtained from (11), and

$$\kappa_i = \frac{\hat{a}_i^2\tilde{E}_i}{\sigma_i^2}, \tag{16}$$

with $\breve{E}_i$ being defined in (6). In other words, the optimal combining technique estimates the time-delay as a weighted average of the time-delays of different branches, where the weights are chosen as proportional to the multiplication of the SNR estimate, $E_i\hat{a}_i^2/\sigma_i^2$, and $\breve{E}_i/E_i$. Since $\breve{E}_i$ is defined as the energy of the first derivative of $s_i(t)$ as in (6), $\breve{E}_i/E_i$ can be expressed, using Parseval's relation, as $\breve{E}_i/E_i = 4\pi^2\beta_i^2$, where β_i is the effective bandwidth of $s_i(t)$, which is defined as [30]

$$\beta_i^2 = \frac{1}{E_i}\int_{-\infty}^{\infty} f^2 |S_i(f)|^2 \mathrm{d}f, \tag{17}$$

with $S_i(f)$ denoting the Fourier transform of $s_i(t)$. Therefore, it is concluded that the optimal combining technique assigns a weight to the time-delay estimate of a given branch in proportion to the product of the SNR estimate and the effective bandwidth related to that branch. The intuition behind this combining technique is the fact that signals with larger effective bandwidths and/or larger SNRs facilitate more accurate time-delay estimation [30]; hence, their weights should be larger in the combining process. This intuition is verified theoretically in Section 5.

4.2.2. SNR Combining

The second technique combines the time-delay estimates in the first step according to the SNR estimates at the respective branches. In other words, the time-delay estimate is obtained as

$$\hat{\tau} = \frac{\sum_{i=1}^{K} \gamma_i \hat{\tau}_i}{\sum_{i=1}^{K} \gamma_i}, \tag{18}$$

where

$$\gamma_i = \frac{\hat{a}_i^2 E_i}{\sigma_i^2}. \tag{19}$$

Note that γ_i defines the SNR estimate at branch i. In other words, this technique considers only the SNR estimates at the branches in order to determine the combining coefficients, and does not take the signal bandwidths into account.

It is observed from (15)–(19) that the optimal combining and the SNR combining techniques become equivalent if $\tilde{E}_1 / E_1 = \cdots = \tilde{E}_K / E_K$. Since $\tilde{E}_i / E_i = 4\pi^2\beta_i^2$, where β_i is the effective bandwidth defined in (17), the two techniques are equivalent when the effective bandwidths of the signals at different branches are all equal.

4.2.3. Selection Combining-1 (SC-1)

Another technique for obtaining the final time-delay estimate is to determine the "best" branch and to use its estimate as the final time-delay estimate. According to SC-1, the best branch is defined as the one that has the maximum value of $\kappa_i = \hat{a}_i^2 \tilde{E}_i / \sigma_i^2$, for $i = 1, \ldots, K$. In other words, the branch with the maximum multiplication of the SNR estimate and the effective bandwidth is determined as the best branch and its estimate is used as the final one. That is,

$$\hat{\tau} = \hat{\tau}_m, \quad m = \arg \max_{i \in \{1,\ldots,K\}} \left\{ \frac{\hat{a}_i^2 \tilde{E}_i}{\sigma_i^2} \right\}, \tag{20}$$

where $\hat{\tau}_m$ represents the time-delay estimate at the m th branch.

4.2.4. Selection Combining-2 (SC-2)

Similar to SC-1, SC-2 selects the "best" branch and uses its estimate as the final time-delay estimate. However, according to SC-2, the best branch is defined as the one with the maximum SNR. Therefore, the time-delay estimate is obtained as follows according to SC-2:

$$\hat{\tau} = \hat{\tau}_m, \quad m = \arg \max_{i \in \{1,\ldots,K\}} \left\{ \frac{\hat{a}_i^2 E_i}{\sigma_i^2} \right\}, \tag{21}$$

where $\hat{\tau}_m$ represents the time-delay estimate at the m th branch.

SC-1 and SC-2 become equivalent when the effective bandwidths of the signals at different branches are all equal.

4.2.5. Equal Combining

The equal combining technique assigns equal weights to the estimates from different branches and obtains the time-delay estimate as follows:

$$\hat{\tau} = \frac{1}{K}\sum_{i=1}^{K}\hat{\tau}_i. \tag{22}$$

Considering the proposed combining techniques above, it is observed that they are similar to diversity combining techniques in communications systems [33]. However, the main difference is the following. The aim is to maximize the SNR or to reduce the probability of symbol error in communications systems [33]; whereas, in the current problem, it is to reduce the MSE of time-delay estimation. In other words, this study considers diversity combining for time-delay estimation, where the diversity results from the dispersed spectrum utilization of the cognitive radio system.

5. ON THE OPTIMALITY OF TWO-STEP TIME-DELAY ESTIMATION

In this section, the asymptotic optimality properties of the two-step time-delay estimators proposed in the previous section are investigated. In order to analyze the performance of the estimators at high SNRs, the result in [24] for time-delay estimation at multiple receive antennas is extended to the scenario in this paper.

Lemma 1.

Consider any linear modulation of the form $s_i(t) = \Sigma_l d_{i,l} p_i(t - lT_i)$, where $d_{i,l}$ denotes the complex data for the l th symbol of signal i, and $p_i(t)$ represents a pulse with duration T_i. Assume that

$\int_{-\infty}^{\infty} s_i'(t-\tau)s_i^*(t-\tau)dt = 0$, for $i = 1, \ldots, K$, then, for the signal model in (1), the delay estimate in (11) and the channel amplitude estimate in (12) can be modeled, at high SNR, as

$$\hat{\tau}_i = \tau + \nu_i, \tag{23}$$

$$\hat{a}_i = a_i + \eta_i, \tag{24}$$

for $i = 1, \ldots, K$, where ν_i and η_i are independent zero mean Gaussian random variables with variances $\sigma_i^2 / (\tilde{E}_i a_i^2)$ and σ_i^2 / E_i, respectively. In addition, ν_i and ν_j (η_i and η_j) are independent for $i \neq j$.

Proof.

The proof uses the derivations in [23] in order to extend Lemma 1 in [24] to the cases with CFO. At high SNRs, the ML estimate $\hat{\boldsymbol{\theta}}_i$ of $\boldsymbol{\theta}_i = [\tau \; a_i \; \phi_i \; \omega_i]$ in (11) and (12) is approximately distributed as a jointly Gaussian random variable with the mean being equal to $\boldsymbol{\theta}_i$ and the covariance matrix being given by the inverse of the Fisher information matrix (FIM) for observation $r_i(t)$ in (1) over $[0, T]$. Then, the results in [23] can be used to show that, under the conditions in the lemma, the first 2×2 block of the covariance matrix can be obtained as $\text{diag}\{\sigma_i^2 / (\tilde{E}_i a_i^2), \sigma_i^2 / E_i\}$. Therefore, $\hat{\tau}_i$ and $\hat{a}_i$ can be modeled as in (23) and (24). In addition, since the noise components at different branches are independent, the estimates are independent for different branches.

Based on Lemma 1, the asymptotic unbiasedness properties of the estimators in Section 4 can be verified. First, it is observed from Lemma 1 that $\mathrm{E}\{\hat{\tau}_i\} = \tau$. Considering the optimal combining technique in (15) as an example, the unbiasedness property can be shown as

$$\begin{aligned} \mathrm{E}\{\hat{\tau} \mid \hat{a}_1, \ldots, \hat{a}_K\} &= \frac{\sum_{i=1}^{K} \kappa_i \mathrm{E}\{\hat{\tau}_i \mid \hat{a}_1, \ldots, \hat{a}_K\}}{\sum_{i=1}^{K} \kappa_i} \\ &= \frac{\sum_{i=1}^{K} \kappa_i \mathrm{E}\{\hat{\tau}_i \mid \hat{a}_i\}}{\sum_{i=1}^{K} \kappa_i} = \tau, \end{aligned} \tag{25}$$

where $\kappa_i = \hat{a}_i^2 \tilde{E}_i / \sigma_i^2$. Since $E\{\hat{\tau} \mid \hat{a}_1, \ldots, \hat{a}_K\}$ does not depend on $\hat{a}_1, \ldots, \hat{a}_K$, $E\{\hat{\tau}\} = E\{E\{\hat{\tau} \mid \hat{a}_1, \ldots, \hat{a}_K\}\} = \tau$. In other words, since for each specific value of $\hat{a}_i$, $\hat{\tau}_i$ is unbiased ($i = 1, \ldots, K$), the weighted average of $\hat{\tau}_1, \ldots, \hat{\tau}_K$ is also unbiased. Similar arguments can be used to show that all the two-step estimators described in Section 4 are asymptotically unbiased.

Regarding the variance of the estimators, it can be shown that the optimal combining technique has a variance that is approximately equal to the CRLB at high SNRs (in fact, this is the main reason why this combining technique is called *optimal*). To that aim, the conditional variance of $\hat{\tau}$ in (15) given $\hat{a}_1, \ldots, \hat{a}_K$ is obtained as follows:

$$\mathrm{Var}\{\hat{\tau} \mid \hat{a}_1, \ldots, \hat{a}_K\} = \frac{\sum_{i=1}^{K} \kappa_i^2 \, \mathrm{Var}\{\hat{\tau}_i \mid \hat{a}_1, \ldots, \hat{a}_K\}}{\left(\sum_{i=1}^{K} \kappa_i\right)^2}, \tag{26}$$

where the independence of the time-delay estimates is used to obtain the result (cf. Lemma 1). Since $\mathrm{Var}\{\hat{\tau}_i \mid \hat{a}_1, \ldots, \hat{a}_K\} = \mathrm{Var}\{\hat{\tau}_i \mid \hat{a}_i\} = \sigma_i^2 / (\tilde{E}_i a_i^2)$ from Lemma 1 and $\kappa_i = \hat{a}_i^2 \tilde{E}_i / \sigma_i^2$, (26) can be expressed as

$$\begin{aligned}\mathrm{Var}\{\hat{\tau} \mid \hat{a}_1, \ldots, \hat{a}_K\} &= \frac{\sum_{i=1}^{K} \left(\hat{a}_i^4 \tilde{E}_i^2 / \sigma_i^4\right)\left(\sigma_i^2 / \left(\tilde{E}_i a_i^2\right)\right)}{\left(\sum_{i=1}^{K} \left(\hat{a}_i^2 \tilde{E}_i / \sigma_i^2\right)\right)^2} \\ &= \sum_{i=1}^{K} \frac{\hat{a}_i^4 \tilde{E}_i}{a_i^2 \sigma_i^2} \left(\sum_{i=1}^{K} \frac{\hat{a}_i^2 \tilde{E}_i}{\sigma_i^2}\right)^{-2}.\end{aligned} \tag{27}$$

Lemma 1 states that at high SNRs, $\hat{a}_i$ is distributed as a Gaussian random variable with mean a_i and variance σ_i^2 / E_i . Therefore, for sufficiently large values of $E_i / \sigma_i^2, \ldots, E_K / \sigma_K^2$, (27) can be approximated by

$$\mathrm{Var}\{\hat{\tau} \mid \hat{a}_1, \ldots, \hat{a}_K\} \approx \left(\sum_{i=1}^{K} \frac{\tilde{E}_i a_i^2}{\sigma_i^2}\right)^{-1}, \tag{28}$$

which is equal to CRLB expression in (8). Therefore, the optimal combining technique in (15) results in an approximately optimal estimator at high SNRs.

The variances of the other combining techniques in Section 4 can be obtained in a straightforward manner and it can be shown that the asymptotic variances are larger than the CRLB in general. For example, for the SNR combining technique in (18), the conditional variance can be calculated as

$$\mathrm{Var}\{\hat{\tau} \mid \hat{a}_1,\ldots,\hat{a}_K\} = \frac{\sum_{i=1}^{K}(\hat{a}_i^4 E_i^2/\sigma_i^4)\left(\sigma_i^2/\left(\tilde{E}_i a_i^2\right)\right)}{\left(\sum_{i=1}^{K}(\hat{a}_i^2 E_i/\sigma_i^2)\right)^2}$$
$$= \sum_{i=1}^{K}\frac{\hat{a}_i^4 E_i^2}{a_i^2\tilde{E}_i\sigma_i^2}\left(\sum_{i=1}^{K}\frac{\hat{a}_i^2 E_i}{\sigma_i^2}\right)^{-2}, \tag{29}$$

which, for sufficiently large SNRs, becomes

$$\mathrm{Var}\{\hat{\tau} \mid \hat{a}_1,\ldots,\hat{a}_K\} \approx \sum_{i=1}^{K}\frac{a_i^2 E_i^2}{\tilde{E}_i\sigma_i^2}\left(\sum_{i=1}^{K}\frac{a_i^2 E_i}{\sigma_i^2}\right)^{-2}. \tag{30}$$

Then, from the Cauchy-Schwarz inequality, the following condition is obtained:

$$\mathrm{Var}\{\hat{\tau} \mid \hat{a}_1,\ldots,\hat{a}_K\} \approx \frac{\sum_{i=1}^{K}\left(a_i^2 E_i^2/\left(\tilde{E}_i\sigma_i^2\right)\right)}{\left(\sum_{i=1}^{K}\left(a_i E_i/\left(\sigma_i\sqrt{\tilde{E}_i}\right)\right)\left(a_i\sqrt{\tilde{E}_i}/\sigma_i\right)\right)^2}$$
$$\geq \frac{\sum_{i=1}^{K}\left(a_i^2 E_i^2/\left(\tilde{E}_i\sigma_i^2\right)\right)}{\sum_{i=1}^{K}\left(a_i^2 E_i^2/\left(\tilde{E}_i\sigma_i^2\right)\right)\sum_{i=1}^{K}\left(a_i^2\tilde{E}_i/\sigma_i^2\right)}$$
$$= \mathrm{CRLB}, \tag{31}$$

which holds with equality if and only if $\tilde{E}_1/E_1 = \cdots = \tilde{E}_K/E_K$ (or, $\beta_1 = \cdots = \beta_K$). In fact, under that condition, the optimal combining and the SNR combining techniques become identical as mentioned in Section 4, since $\kappa_i = \hat{a}_i^2\tilde{E}_i/\sigma_i^2 = (\tilde{E}_i/E_i)(E_i\hat{a}_i^2/\sigma_i^2) = (\tilde{E}_i/E_i)\gamma_i$ (cf. (16) and (19)). In other words, when the effective bandwidths of the signals at different branches are not equal, the asymptotic variance of the SNR combining technique is strictly larger than the CRLB.

Regarding the selection combining approaches in (20) and (21), similar conclusions as for the diversity combining techniques in communications systems can be made [33]. Specifically, SC-1 and SC-2 perform worse than the optimal combining and the SNR combining techniques, respectively, in general. However, when the estimate of a branch is significantly more accurate than the others, the performance of the selection combining approach can get very close to the optimal combining or the SNR combining technique. However, when the branches have similar estimation accuracies, the selection combining techniques can perform significantly worse. The conditional variances of the selection combining techniques can be approximated at high SNR as

$$\mathrm{Var}\{\hat{\tau} \mid \hat{a}_1,\ldots,\hat{a}_K\} \approx \min\left\{\frac{\sigma_1^2}{\tilde{E}_1 a_1^2},\ldots,\frac{\sigma_K^2}{\tilde{E}_K a_K^2}\right\}, \tag{32}$$

for SC-1, and

$$\mathrm{Var}\{\hat{\tau} \mid \hat{a}_1,\ldots,\hat{a}_K\} \approx \frac{E_m}{\tilde{E}_m}\min\left\{\frac{\sigma_1^2}{E_1 a_1^2},\ldots,\frac{\sigma_K^2}{E_K a_K^2}\right\}, \tag{33}$$

for SC-2, where $m = \arg\min_{i \in \{1,\ldots,K\}}\{\sigma_i^2 / (E_i \hat{a}_i^2)\}$. From (32) and (33), it is observed that if $\tilde{E}_1/E_1 = \cdots = \tilde{E}_K/E_K$ ($\beta_1 = \cdots = \beta_K$), then the asymptotic variances of the SC-1 and SC-2 techniques become equivalent.

Finally, for the equal combining technique, the variance can be obtained from (22) as

$$\mathrm{Var}\{\hat{\tau}\} = \frac{1}{K^2}\sum_{i=1}^{K}\frac{\sigma_i^2}{\tilde{E}_i a_i^2}. \tag{34}$$

In general, the equal combining technique is expected to have the worst performance since it does not make use of any information about the SNR or the signal bandwidths in the estimation of the time-delay.

6. SIMULATION RESULTS

In this section, simulations are performed in order to evaluate the performance of the proposed time-delay estimators and compare them with each other and against the CRLBs. The signal $s_i(t)$ in (1) corresponding to each branch is modeled by the Gaussian doublet given by

$$s_i(t) = A_i\left(1 - \frac{4\pi(t - 1.25\,\zeta_i)^2}{\zeta_i^2}\right)e^{-2\pi(t-1.25\,\zeta_i)^2/\zeta_i^2}, \tag{35}$$

where A_i and ζ_i are the parameters that are used to adjust the pulse energy and the pulse width, respectively. The bandwidth of $s_i(t)$ in (35) can approximately be expressed as $B_i \approx 1/(2.5\zeta_i)$ [29]. For the following simulations, A_i values are adjusted to generate unit-energy pulses.

For all the simulations, the spectral densities of the noise at different branches are assumed to be equal; that is, $\sigma_i^2 = \sigma^2$ for $i = 1, \ldots, K$. In addition, the SNR of the system is defined with respect to the total energy of the signals at different branches; that is, $\mathrm{SNR} = 10\log_{10}(\Sigma_{i=1}^{K} E_i / (2\sigma^2))$.

In assessing the root-mean-squared errors (RMSEs) of the different estimators, a Rayleigh fading channel is assumed. Namely, the channel coefficient $\alpha_i = a_i e^{j\phi_i}$ in (1) is modeled as a_i being a Rayleigh distributed random variable and ϕ_i being uniformly distributed in $[0, 2\pi)$. Also, the same average power is assumed for all the bands; namely, $\mathrm{E}\{|\alpha_i|^2\} = 1$ is used. The time-delay τ in (1) is uniformly distributed over the observation interval. In addition, it is assumed that there is no CFO in the system.

First, the performance of the proposed estimators is evaluated with respect to the SNR for a system with $K = 3$, $B_1 = 200$ kHz, $B_2 = 100$ kHz, and $B_3 = 400$ kHz. The results in Figure 4 indicate that the optimal combining technique has the best performance as expected from the theoretical analysis, and SC-1, which estimates the delay according to (20), has performance close to that of the optimal combining technique. The SNR combining and SC-2 techniques have worse performance than that of the optimal and SC-1 techniques, respectively. In addition, SC-1 has better performance than that of the SNR combining technique in this scenario, which indicates that selecting the delay estimate corresponding to the largest $\breve{E}_i \hat{a}_i^2 / \sigma_i^2$ value is closer to optimal than combining the delay estimates of the different branches according to the SNR combining criterion in (18) for the considered scenario. The main reason for this is related to the large variability of the channel amplitudes due to the nature of the Rayleigh distribution. Since the channel amplitude levels are expected to be quite different for most of the time, using the delay estimate of the best one yields a more reliable estimate than combining the delay estimates according to the *suboptimal* SNR combining technique (since the signal bandwidths are different, the SNR combining technique is suboptimal as studied in Section 5). Regarding the equal combining technique, it has significantly worse performance than the others, since it combines all the delay estimates equally. Since the delay estimates of some branches can have very large errors due to fading, the RMSEs of the equal combining technique become significantly larger. For example, when converted to distance estimates, an RMSE of about 120meters is achieved by this technique, whereas the optimal combining technique results in an RMSE of less than 15 meters. Finally, it is observed that the performance of the optimal combining technique gets very close to the CRLB at high SNRs, which is expected from the asymptotic arguments in Section 5.

In addition, the equivalence of the optimal combining and the SNR combining techniques and that of SC-1 and SC-2 are illustrated in Figure 6, where $K = 2$ and $B_1 = B_2 = 400$ kHz are used. In other words, the signal consists of two dispersed bands with 400 kHz bandwidths, and in each band, the same signal described by (35) is used. Therefore, $\breve{E}_1 / E_1 = \breve{E}_2 / E_2$ is satisfied, which results in the equivalence of the optimal combining and the SNR combining techniques, as well as that of the SC-1 and SC-2 techniques, as discussed in Section 4. Also, since there are only two bands ($K = 2$), the selection combining techniques get very close to the optimal combining and the SNR combining techniques.

Next, similar performance comparisons are performed for a signal with $K = 2$, $B_1 = 100$ kHz, and $B_2 = 400$ kHz, as shown in Figure 5. Again similar observations as for Figure 4 are made. In addition, since there are only two bands ($K = 2$) and the signal bandwidths are quite different, the selection combining techniques, SC-1 and SC-2, get very close to the optimal combining and the SNR combining techniques, respectively.

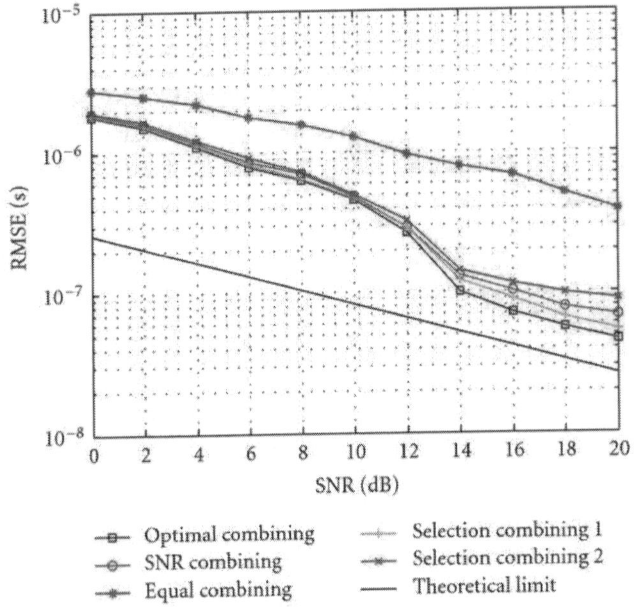

Figure 4 RMSE versus SNR for the proposed algorithms, and the theoretical limit (CRLB). The signal occupies three dispersed bands with bandwidths B_1 = 200 kHz, B_2 = 100 kHz and B_3 = 400 kHz.

In Figure 7, the RMSEs of the proposed estimators are plotted against the number of bands, where each band is assumed to have 100 kHz bandwidth. The spectral densities are set to $\sigma_i^2 = \sigma^2 = 0.1$ for all i. Since the same signals are used in each band, the optimal combining and the SNR combining techniques become identical; hence, only one of them is marked in the figure. Similarly, since SC-1 and SC-2 are identical in this scenario, they are referred to as "selection combining" in the figure. It is observed from Figure 7 that the optimal combining has better performance than the selection combining and the equal combining techniques. In addition, as the number of bands increases, the amount of reduction in the RMSE per additional band decreases (i.e., diminishing return). In fact, the selection combining technique seems to converge to an almost constant value for large numbers of bands. This is intuitive since the selection combining technique always uses the estimate from one of the branches; hence, in the presence of a sufficiently large number of bands, additional bands do not cause a significant increase in the diversity. On the other hand, the optimal combining technique has a slope that is quite similar to that of the CRLB; that is, it makes an efficient use of the frequency diversity.

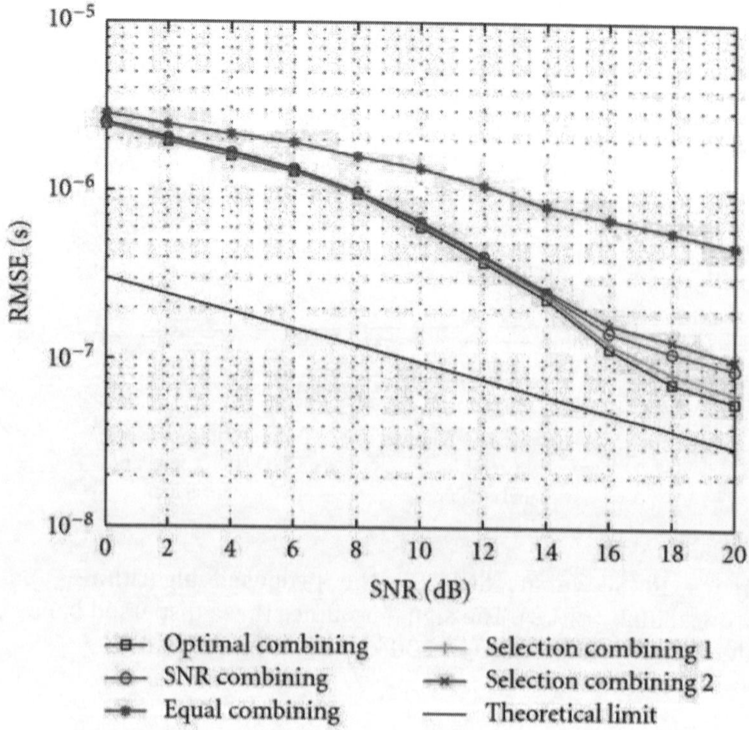

Figure 5 RMSE versus SNR for the proposed algorithms, and the theoretical limit (CRLB). The signal occupies two dispersed bands with bandwidths $B_1 = 100$ kHz and $B_2 = 400$ kHz.

Finally, the performance of the proposed algorithms is investigated in the presence of CFO in Figure 8. The CFOs at different branches are modeled by independent uniform random variables over $[-100, 100]$ Hz, and the RMSEs are obtained for the system parameters that are considered for Figure 5. Again similar observations as for Figures 4 and 5 are made. In addition, the comparison of Figures 5 and 8 reveals that the RMSE values slightly increase in the presence of CFOs, although the theoretical limit stays the same [23].

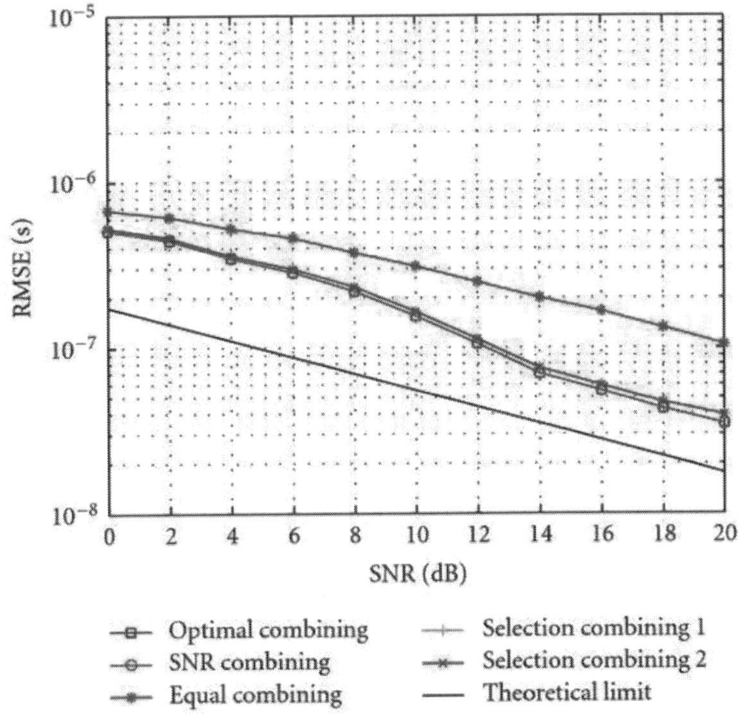

Figure 6 RMSE versus SNR for the proposed algorithms, and the theoretical limit (CRLB). The signal occupies two dispersed bands with equal bandwidths of 400 kHz.

7. CONCLUDING REMARKS

Time-delay estimation for dispersed spectrum cognitive radio systems has been studied. After the investigation of the ML estimator and the CRLBs, a two-step approach has been proposed to obtain accurate time-delay estimates with reasonable computational complexity. In the first step of the proposed approach, an ML estimator is used at each branch of the receiver in order to estimate the unknown parameters of the received signal at that branch. Then, in the second step, a number of diversity combining approaches have been studied. In the optimal combining technique, both the SNRs and the bandwidths of the signals at different branches are considered to obtain the time-delay estimate; whereas the SNR combining technique obtains the time-delay estimate according to the estimated SNR values only. In addition, two selection combining techniques, as well as the equal combining technique, have been investigated. It has been shown that the optimal combining technique can approximate the CRLB at high SNRs; whereas the equal combining technique has the worst performance since it does not make use of any information about signal bandwidths and/or the SNRs. Simulation results have been presented to verify the theoretical analysis.

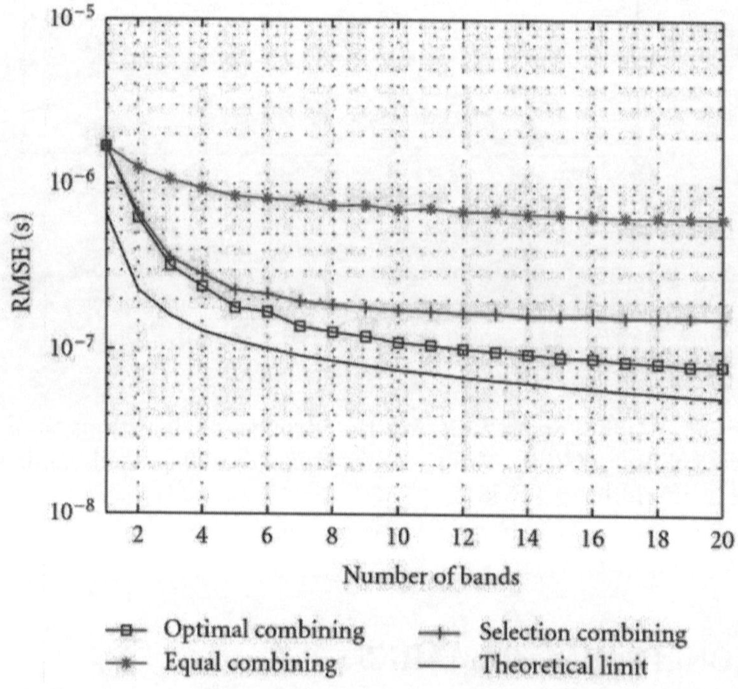

Figure 7 RMSE versus the number of bands for the proposed algorithms, and the theoretical limit (CRLB). Each band occupies 100 kHzz, and $\sigma_i^2 = 0.1$ for all *i*.

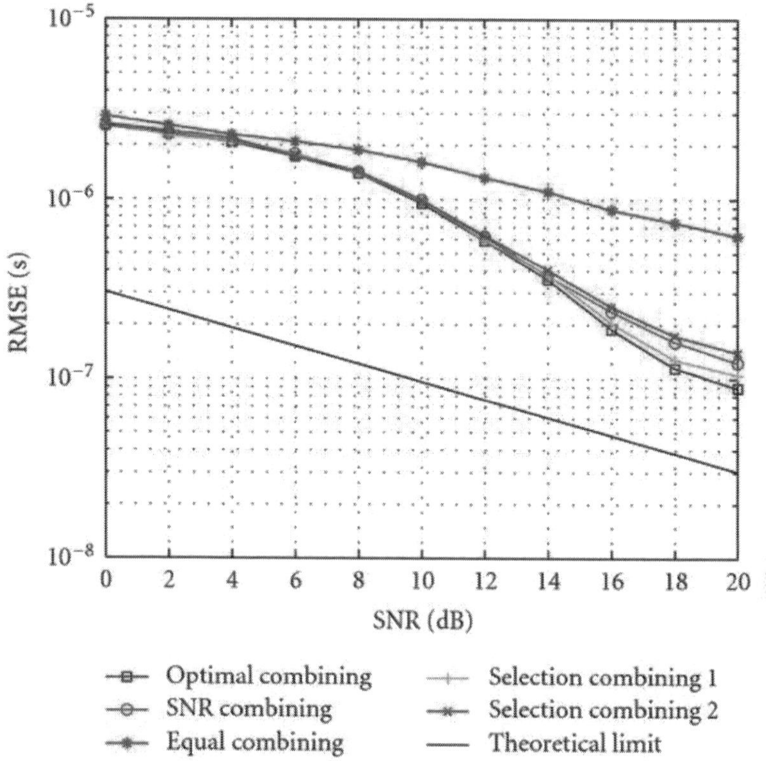

Figure 8 RMSE versus SNR for the proposed algorithms, and the theoretical limit (CRLB) in the presence of CFO. The signal occupies two dispersed bands with bandwidths $B_1 = 100$ kHz and $B_2 = 400$ kHz.

ACKNOWLEDGMENTS

F. Kocak and S. Gezici wish to acknowledge the activity of the Network of Excellence in Wireless COMmunications NEWCOM++ of the European Commission (Contract no. 216715) that motivated this work. This work was supported in part by the WiMAGIC project of the EC Seventh Framework Programme (FP7) under Grant agreement no. 215167, and in part by the U. S. National Science Foundation under CNS-09-05398. H. Celebi and K. A. Qaraqe wish to acknowledge the support of Qatar National Funds for Research (QNRF) and Qatar Telecom (Qtel) during this work.

REFERENCES

1. Mitola J III, Maguire GQ Jr.: Cognitive radio: making software radios more personal. *IEEE Personal Communications* 1999, 6(4):13-18. 10.1109/98.788210

2. Haykin S: Cognitive radio: brain-empowered wireless communications. *IEEE Journal on Selected Areas in Communications* 2005, 23(2):201-220.

3. Quan Z, Cui S, Poor HV, Sayed AH: Collaborative wideband sensing for cognitive radios. *IEEE Signal Processing Magazine* 2008, 25(6):60-73.

4. Zhao Q, Sadler BM: A survey of dynamic spectrum access. *IEEE Signal Processing Magazine* 2007, 24(3):79-89.

5. Devroye N, Vu M, Tarokh V: Cognitive radio networks. *IEEE Signal Processing Magazine* 2008, 25(6):12-23.

6. Scutari G, Palomar DP, Barbarossa S: Cognitive MIMO radio. *IEEE Signal Processing Magazine* 2008, 25(6):46-59.

7. Budiarjo I, Nikookar H, Ligthart LP: Cognitive radio modulation techniques. *IEEE Signal Processing Magazine* 2008, 25(6):24-34.

8. Arslan H (Ed): *Cognitive Radio, Software Defined Radio, and Adaptive Wireless Systems*. Springer, London, UK; 2007.

9. Neel JO: *Analysis and design of cognitive radio networks and distributed radio resource management algorithms, Ph.D. dissertation*. Virginia Polytechnic Institute and State University, Blacksburg, Va, USA; 2006.

10. Celebi H, Arslan H: Enabling location and environment awareness in cognitive radios. *Computer Communications* 2008, 31(6):1114-1125. 10.1016/j.comcom.2008.01.006

11. Federal Communications Commission (FCC) : Facilitating opportunities for flexible, efficient, and reliable spectrum use employing cognitive radio technologies. *ET Docket* 2005., (03-108):

12. Wellens M, Wu J, Mahonen P: Evaluation of spectrum occupancy in indoor and outdoor scenario in the context of cognitive

radio.*Proceedings of the 2nd International Conference on Cognitive Radio Oriented Wireless Networks and Communications (CrownCom '07), August 2007, Orlando, Fla, USA* 420-427.

13. Haykin S: Cognitive radar: a way of the future. *IEEE Signal Processing Magazine* 2006, 23(1):30-40.

14. Zhao Y, Reed JH, Mao S, Bae KK: Overhead analysis for radio environment mapenabled cognitive radio networks. *Proceedings of the 1st IEEE Workshop on Networking Technologies for Software Defined Radio Networks (SDR '06), September 2006, Reston, Va, USA* 18-25.

15. Celebi H, Arslan H: Utilization of location information in cognitive wireless networks. *IEEE Wireless Communications* 2007, 14(4):6-13.

16. Yarkan S, Arslan H: Exploiting location awareness toward improved wireless system design in cognitive radio. *IEEE Communications Magazine* 2008, 46(1):128-136.

17. Celebi H, Arslan H: Cognitive positioning systems. *IEEE Transactions on Wireless Communications* 2007, 6(12):4475-4483.

18. Mahonen P, Petrova M, Riihijarvi J: Applications of topology information for cognitive radios and networks. *Proceedings of the 2nd IEEE International Symposium on New Frontiers in Dynamic Spectrum Access Networks (DySpan '07), April 2007, Dublin, Ireland* 103-114.

19. Celebi H, Arslan H: Adaptive positioning systems for cognitive radios. *Proceedings of the 2nd IEEE International Symposium on New Frontiers in Dynamic Spectrum Access Networks (DySpan '07), April 2007, Dublin, Ireland* 78-84.

20. Celebi H, Arslan H: Utilization of location information in cognitive wireless networks. *IEEE Wireless Communications* 2007, 14(4):6-13.

21. Gezici S: A survey on wireless position estimation. *Wireless Personal Communications* 2008, 44(3):263-282. 10.1007/s11277-007-9375-z

22. Gezici S, Tian Z, Giannakis GB, *et al.*: Localization via ultra-wideband radios: a look at positioning aspects of future sensor networks. *IEEE Signal Processing Magazine* 2005, 22(4):70-84.

23. Gezici S, Celebi H, Poor HV, Arslan H: Fundamental limits on time delay estimation in dispersed spectrum cognitive radio systems. *IEEE Transactions on Wireless Communications* 2009, 8(1):78-83.

24. Gezici S, Sahinoglu Z: Ranging in a single-input multiple-output (SIMO) system. *IEEE Communications Letters* 2008, 12(3):197-199.

25. Williams C, McLaughlin S, Beach MA: Exploiting multiple antennas for synchronization. *IEEE Transactions on Vehicular Technology* 2009, 58(2):773-787.

26. Weiss TA, Jondral FK: Spectrum pooling: an innovative strategy for the enhancement of spectrum efficiency. *IEEE Communications Magazine* 2004, 42(3):S8-S14.

27. Brandes S, Cosovic I, Schnell M: Reduction of out-of-band radiation in OFDM systems by insertion of cancellation carriers. *IEEE Communications Letters* 2006, 10(6):420-422. 10.1109/ LCOMM. 2006.1638602

28. Mahmoud HA, Arslan H: Sidelobe suppression in OFDM-based spectrum sharing systems using adaptive symbol transition. *IEEE Communications Letters* 2008, 12(2):133-135.

29. Sahinoglu Z, Gezici S, Guvenc I: *Ultra-Wideband Positioning Systems: Theoretical Limits, Ranging Algorithms, and Protocols*. Cambridge University Press, Cambridge, UK; 2008.

30. Poor HV: *An Introduction to Signal Detection and Estimation*. Springer, New York, NY, USA; 1994.

31. Gezici S, Sahinoglu Z, Molisch AF, Kobayashi H, Poor HV: Two-step time of arrival estimation for pulse-based ultra-wideband systems. *EURASIP Journal on Advances in Signal Processing* 2008, 2008:-11.

32. Kim T-H, Park I-C: Two-step approach for coarse time synchronization and frequency offset estimation for IEEE 802.16D systems.*Proceedings of IEEE Workshop on Signal Processing Systems (SiPS '07), October 2007, Shanghai, China* 193-198.

33. Goldsmith A: *Wireless Communications*. Cambridge University Press, Cambridge, UK; 2005.

6

DELAY/DISRUPTION TOLERANT NETWORK-BASED MESSAGE FORWARDING FOR A RIVER POLLUTION MONITORING WIRELESS SENSOR NETWORK APPLICATION

Carlos Velásquez-Villada * and Yezid Donoso

Systems and Computing Engineering Department, Universidad de los Andes, 111711 Bogotá D.C., Colombia

ABSTRACT

Communications from remote areas that may be of interest is still a problem. Many innovative projects applied to remote sites face communications difficulties. The GOLDFISH project was an EU-funded project for river pollution monitoring in developing countries. It had several sensor clusters, with floating WiFi antennas, deployed along a downstream river's course. Sensor clusters sent messages to a Gateway installed on the riverbank. This gateway sent the messages, through a backhaul technology, to an Internet server where data was aggregated over a map. The communication challenge in this scenario was produced by the antennas' movement and network backhaul availability. Since the antennas were floating on the river, communications could be disrupted at any time. Also, 2G/3G availability near the river was not constant. For non-real-time applications, we propose a Delay/Disruption Tolerant Network (DTN)-based solution where all nodes have persistent storage capabilities and DTN protocols to be able to wait minutes or hours to transmit. A mechanical backhaul will periodically visit the river bank where the gateway is installed and it will automatically collect sensor data to be carried to an Internet-covered spot. The proposed forwarding protocol delivers around 98% of the messages for this scenario, performing better than other well-known DTN routing protocols.

KEYWORDS

disruption-tolerant; delay-tolerant; WSN; delivery probability; opportunistic forwarding; pollution monitoring; rural telecommunications

1. INTRODUCTION

The "GOLDFISH—Detection of Watercourse Contamination in Developing Countries using Sensor Networks" project was an Information and Communication Technologies (ICT) project funded from 2012 to 2015 by the European Union (EU). It was developed by a consortium of 11 universities and organizations from Poland, France, Sweden, Croatia, Slovakia, Colombia, and Bolivia. The GOLDFISH project worked with the Raspberry Pi platform to integrate all its components and developed the sensors technology, the communication modules, the Central Management and Monitoring platform, and the communication algorithms. In the project's extension, an energy generation technology and a pollutant locator application were added [1].

The GOLDFISH project was designed, as can be seen in Figure 1, to sense river pollution with some embedded sensors, and to send these data to a Gateway (GW) on the river's shore. The GW sent sensor data to the central Management and Monitoring Station (MMS) via VPN through the Mobile network. The communication between the nodes in the river and between the nodes and the GW was achieved via WiFi (IEEE 802.11b) with a floating antenna incorporated in the sensor node design (see Figure 2). The GW had 2G/3G communication capabilities and it sent the sensor data through this channel. Once the MMS had the sensor data, it could show anomalies or alarms for higher than expected levels of pollutants. The system was tested in Europe and in Colombia. However, for tests in Colombia, the system had communication failures. The communication from the GW to the MMS did not work. There was a cellular antenna near the test site (it can be seen inFigure 2), but the data throughput varied between 20 and 95 kbps, and the data delay varied from 450 ms to 2.12 s to reach a server in Bogotá D.C. (100 km from the test site). In addition, the VPN to the MMS had to be created with a server in Warsaw. After two days of testing, this communication part of the project (GW to MMS) did not work. These problems were probably due to the lack of availability of 3G channels at the test site and the high delays experienced in the 2G channel available. This might have caused the VPN request to expire before connecting with the MMS or before the data could be sent to the MMS.

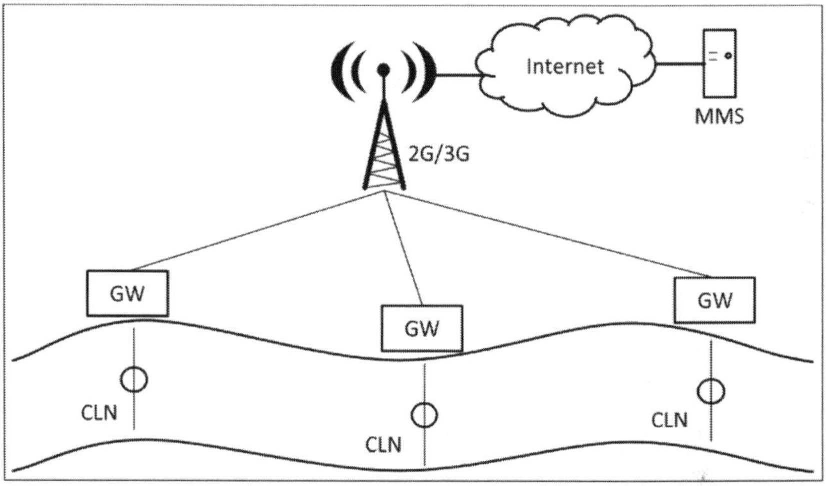

Figure 1. GOLDFISH project description. Sensor Nodes (CLN), Management and Monitoring Station (MMS), and Gateways (GW) are depicted.

To solve the GOLDFISH project communication problem and to deal with even more restricted communication capabilities that can be present on more remote sites, we are proposing a Delay/Disruption Tolerant Network (DTN) that can use local transportation facilities to send and receive messages to Internet servers for non real-time and non-time-critical applications. The main idea is to send the sensor node information to the GW when possible or to a neighboring sensor node if it has a better probability to deliver the information to the GW. Sensor nodes (CLN) and the GW will store all data until they can reach the next hop en route to their destination. For the GW, it will be a mobile node that will transport these data to an Internet connected site, where they will be sent to the MMS (See Figure 3). Sensor nodes can communicate with each other and with the GW and learn who are their neighbors, how often they meet, and calculate the delivery probabilities through other nodes, so they can make decisions about which node will be the next hop in their transmission. The GW will receive all messages from the sensor nodes and will wait until a mobile node (MAP) comes by and collects the data. These MAPs will be mechanical vehicles that will, eventually or periodically, visit the site and contact the GW. When the GW encounters a MAP it will calculate the MAP's delivery probability, and will decide if it gives the MAP all the data or if it should wait for the visit of another MAP with a better delivery probability. The GW can calculate these delivery probabilities based on historical information of contacts and the delivery probabilities given by these same MAPs. More details about the solution will be given in Section 3.

Figure 2. GOLDFISH project running in Colombia, 2015.

Section 2 will present some previous and related work for this kind of network. The proposal mentioned above, as well as its detailed algorithm, are presented in Section 3. Simulation results will be shown and analyzed in Section 4. Finally, conclusions will be presented in Section 5.

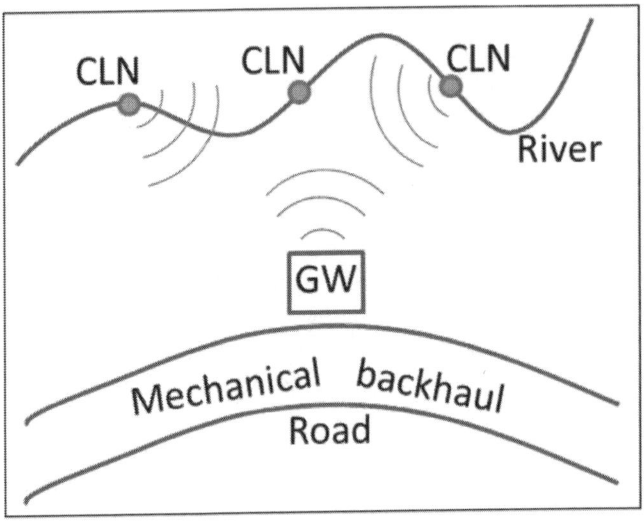

Figure 3. Modified GOLDFISH project. Cluster Nodes (CLN) and Gateway (GW) are depicted.

2. PREVIOUS AND RELATED WORKS

In this project we are looking for a connectivity solution in a remote, usually disconnected site, near a river in underdeveloped countries, to connect a Wireless Sensor Network (WSN) solution for pollution monitoring. A possible solution for this scenario are Delay/Disruption Tolerant Networks (DTN). DTNs have been used in rural applications for several years. Although DTNs started as an alternative to spatial communications led by NASA, they have expanded to several applications in Earth-based communications; for example, Wireless Sensor Networks, Remote and Extreme Environments Monitoring, and Rural Communications. The following subsections describe the DTN Architecture, the Bundle Protocol, some routing and forwarding algorithms, and some rural connectivity technologies.

2.1. DTN Architecture and Bundle Protocol

The DTN Architecture [2] of 2007 describes an architecture for networks with special conditions where traditional protocols (TCP/IP) will not work. This architecture defines a message-oriented layer over the traditional transport layer, the Bundle Layer. The Bundle Layer establishes a store-and-forward kind of network with persistent storage as a measure against network disconnections, and uses Uniform Resource Identifiers (URI) [3] as a naming scheme. Applications can deliver and read Application Data Units (ADUs) to and from the Bundle Layer, which will encapsulate them into Bundles. These Bundles will be stored in persistent storage, and the responsibility of delivery can be transferred with the Bundle, creating a kind of distributed end-to-end reliability. The DTN Architecture allows unicast, anycast, and multicast communications, and it can also use priorities. Finally, the DTN Architecture can take advantage of continuous, scheduled, predicted, and opportunistic connectivity.

The Bundle Protocol [4] defines the exchange of messages (Bundles) between DTN nodes. The Bundle Protocol works as an application in the traditional networking model, since it seats on top of the transport layer, communicating DTN applications between DTN nodes. Its inner working looks like a communication between a couple of neighboring nodes for normal protocols, but this communication can include the transfer of custody and responsibility of delivery of a message. DTN nodes, then, can store Bundles until they have the opportunity to send them to their destination or to another node that can deliver them. In normally disconnected or high-delay networks, this storage can lasts for minutes, hours or days. The Bundle Protocol is independent of networking protocols in the layers below its implementation, and applications should be DTN-compliant or at least know what to do when a message arrives long after its request.

2.2. Rural Connectivity Technologies

Recently, there have been many initiatives to connect the unconnected half of the world. These unconnected citizens are, primarily, located in rural areas in developing countries. Main players in the IT world, like Google, Facebook and Intel, are heading these initiatives. In the next paragraphs we will review a couple of their projects and a DTN-Based rural connectivity technology, the DakNet project.

Google's Loon project [5] is a Google initiative to bring Internet to underdeveloped and remote regions using stratospheric, solar-powered balloons. It uses base stations on Earth, below the balloon's trajectory, and it creates a mesh network between the balloons to transfer messages to these ground base stations with Internet connectivity to transmit replies and requests. Users can connect to the balloons via LTE (Long Term Evolution) devices through a local carrier's infrastructure.

Facebook's Internet.org [6] uses local infrastructure, along with satellite communications and unmanned autonomous vehicles (UAV), to deliver a limited version of Internet to rural areas not currently covered or where access to Internet is very expensive. Users need an Internet.org account or to be affiliated with Facebook's local mobile partner and they will have access to Facebook, local news, employment, health, and weather websites.

The main example of DTN use for Rural Connectivity is DakNet [7,8]. DakNet is a technology based on DTNs that uses buses and local transportation infrastructure to transport messages between small and remote towns and Internet. DakNet has been in use for several years in India and Bangladesh and it uses PROPHET [9] (see next subsection for more details on this protocol) as its routing protocol. DakNet establishes small kiosks in rural towns with no Internet connectivity that allow people to send emails and request information offline, and then, using the transportation infrastructure that comes into town, it transports these requests to Internet using the connection at the bus station in the nearest city and delivers their answers when the transport returns to town.

There are alternatives to use DTNs in city environments, like the one proposed in [10]. In this proposal, DTNs can be used to collect sensor data for Smart Cities in an Urban Automation Network (UAN). The nodes can be installed in different parts of the city, and a Garbage truck or a bus will eventually (at least once a day, according to the authors) collect the sensor information. In the paper, DTNs are compared with different alternatives and conclude that they are an efficient solution for sporadic transmissions.

2.3. DTN Routing and Forwarding Algorithms

Epidemic routing [11] is one of the earliest opportunistic routing protocols. In its basic form, a node running epidemic routing replicates the messages it needs to send to every neighbor it meets. Then, each message is replicated by those neighbors to their neighbors, until, eventually, the message reaches its destination. Epidemic routing can guarantee the delivery of all messages if enough resources are available.

However, it can have problems with buffer and bandwidth management, as the messages can quickly exhaust the available capacity.

Spray and Wait routing [12] is an evolution of epidemic routing. A node running Spray and Wait also replicates the messages it needs to send, but it does so in two phases. The first phase, Spray, replicates a limited number of copies using one of two schemes (Source or Binary). The second phase, Wait, just transmits the message when it encounters its destination, stopping the replication process.

BUBBLE routing [13] is a protocol that exploits social features of the nodes communicating in a intermittently connected network. It uses labels to know to which communities or subcommunities the nodes belong to, and it also uses the centrality of the nodes to determine a ranking and take forwarding decisions.

The most known routing protocol for DTNs is PROPHET [9]. PROPHET stands for a Probabilistic Routing Protocol using History of Encounters and Transitivity. A node running PROPHET uses a history of contacts to assign delivery probabilities to nodes it meets and also to nodes its contacts can reach (transitivity). These probabilities decay over time so nodes can predict delivery probabilities to remote nodes.

Authors in [14] propose MPAD (Mobility Prediction-based Adaptive Data gathering protocol), a routing protocol that replicates messages and sends them to high delivery probability nodes. This protocol manages nodes' queues based on the messages' time to live, and it manages forwarding decisions based on direction movement regarding sink position. Authors assume that nodes have GPS or positioning capabilities and that nodes can determine where the sink is.

Other social-based routing protocols for DTNs are SimBet [15] and PeopleRank [16]. These protocols use some measurements or estimations about nodes' centrality, betweenness, similarity, or social distance to rank or qualify neighboring nodes and decide whether or not to pass them the message.

Finally, other hybrid routing protocols are RAPID [17], SMART [18], and LSF [19]. RAPID (Resource Allocation Protocol for Intentional DTN) qualifies neighboring nodes with a utility function and replicates its message to the best qualified nodes. SMART (Selectively MAking pRogress Towards delivery) creates communities of nodes called companions and replicates the messages to the destination companions and, after a time threshold, the nodes will behave like binary Spray-and-

Wait. LSF (Last Seen First spraying) passes several copies of a message (sprays) to a neighboring node with a more recent contact time to the destination.

Table 1 summarizes the classification of these previously described protocols. More routing protocols and other classifications can be found in reviews [20,21,22]. Our algorithm uses ideas from some of the protocols described in this section. It can be considered a hybrid protocol, since it has social, probabilistic, and replication characteristics. In the next section we will give more details about it.

Table 1. Related protocols.

Characteristics	Protocols
Flooding-based	Epidemic [11], Spray and Wait [12]
Social-based	SimBet [15], BUBBLE [13], and PeopleRank [16]
Hybrid protocols	PROPHET [9], MPAD [14] RAPID [17], LSF [19] and SMART [18]

3. A NON-REAL-TIME DTN-BASED FORWARDING ALGORITHM

3.1. A New Multi-Objective Mathematical Model

In this section we introduce a new mathematical model, built over our previous research [23], that will describe the message forwarding from sensor nodes, through Gateways (GWs) and Mobile Access Points (data mules or MAPs), to the Internet Connected Server (ICS) where the sensor data will be sent to the Central Management and Monitoring Platform (MMS). This mathematical model assumes that the communication will only take place from sensor nodes (source of information) to the MMS (the ICS will be the destination).

The mathematical model is presented in Equations (2)–(8). We are proposing a model for a network where their nodes can change their availability at any given time. The availability changes for the sensor nodes come from the fact that the GOLDFISH project uses microwave communication (IEEE 802.11b) and antennas floating over water. The antennas movement by the river flow can cause the water to cover them or tilt them to an angle where communication with the GW is no longer possible. Also, communication from the GW to the MMS is not always possible through the mobile cell network, so we will instead use local transportation means like buses or boats to recover sensor data from the GW and take them to the nearest ICS where it could be transferred to the MMS. The availability of the MAPs will change depending on their physical closeness with the GW. We assume that these vehicles transporting the MAPs will periodically or eventually visit the site where the GW is located. Figure 4 shows a graphical representation of the problem, where the availability probability $a_{ij}(t)$ between each pair of nodes depends on the existence (at any given time) of a physical link between them, and the delivery probability $d_{ij}(t)$ depends on the availability probability and the knowledge of the next node about a path to the destination, see Equation (1). In Equation (1), node k is the destination node, node i is the source node, and node j is an intermediary node. If the node j can deliver the message to node k at time t, the delivery probability $d_{jk}(t)$ will be equal to $a_{jk}(t)$.

$$d_{ij}(t) = a_{ij}(t)d_{jk}(t) \tag{1}$$

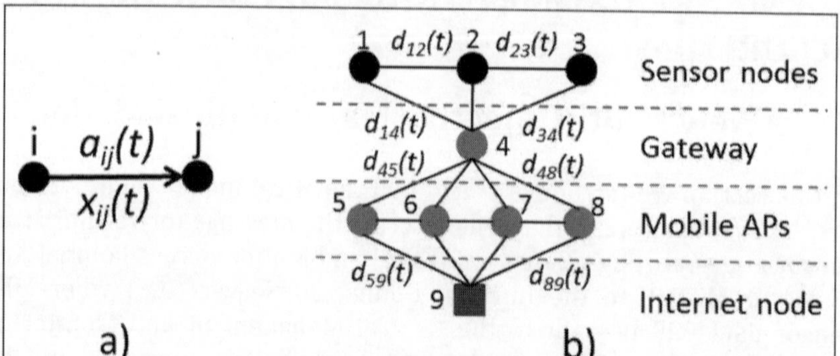

Figure 4. Problem graph. (**a**) Availability probability and data flow from node *i* to node*j*; (**b**) Network graph with delivery probabilities varying through time. AP: Access Point.

Our mathematical model will have a set of nodes (N) where some nodes will be information-originating ones (sensor nodes in the river), others will be transit nodes (GW and MAPs), and one will be the destination node (the ICS in contact with the MMS). Each of these nodes will have a demand or an offer modeled by the parameter $b_i(t)$ (an integer vector) that can change over time. There is also a set of paths (*P*) that represent possible paths over the network through time. Paths are given by possible physical connections, so a path can be created through contacts of several nodes even if they happen at different times. The delivery probabilities $d_{ij}(t)$ are each of the probabilities that a node *i* can reach its destination through a node *j*. They depend on the link availability probability and the next node delivery probability to the destination. Equation (1) describes the delivery probability for a node *i* to reach a node *k* through a node *j*. Links between nodes have a given capacity modeled by c_{ij}, and nodes also have buffers to store data for some time, modeled by c_{ii}. We are assuming these capacities to be constant over time. The decision to send data over a link or store it until later is modeled through the positive integer variable $x_{ij}(t)$. This decision variable models the amount of data that is sent through a link at any given time and it is used to maximize the amount of data flowing through the highest delivery probability paths and minimize the time used to get the data to their destination. Table 2 shows all parameters and sets used in the model. Figure 5 shows the network graph expanded through time.

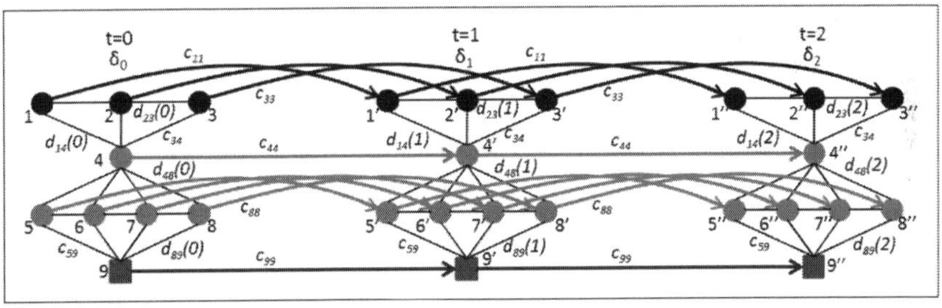

Figure 5. Problem graph through time. Temporal links go from left to right and represent nodes' buffers.

The multiobjective mathematical model maximizes the data flow through the paths with the highest delivery probabilities over time and at the same time it tries to minimize the delivery time for these messages in the network (see Equations (2) and (3). Constraints are described in Equations (4)–(7). Equations (4) and (5) are flow and balance constraints that require demands to be met, so all messages are delivered to their destination through one or several time steps. Equation (4) states that at time zero (0) what a node sends to other neighboring nodes and stores in its buffer should be equal to the offer of the node $(b_i(t))$ plus the messages received from other nodes. Equation (5) is a generalization of Equation (4) for times greater than zero. It takes into account the messages that are stored in a nodes' buffer from the previous time. Equations (6) and (7) are capacity constraints that require that for any given time, information flow over links and stored in buffers must be equal to or below these links and buffers' capacities. Finally, constraint in Equation (8) states that the decision variable $x_{ij}(t)$ is a positive integer defined over all links for all paths.

Table 2. Mathematical model parameters.

Parameter	Definition
N	Set of nodes
E	Set of edges
P	Set of candidate paths
T	Set of discrete time values
c_{ij}	Capacity of edge $(i, j) \in E$
c_{ii}	Storage capacity of node i
$a_{ij}(t)$	Availability Probability of edge $(i, j) \in E$ at time t. $a_{ij}(t) = a_{ji}(t)$
$d_{ij}(t)$	Delivery Probability of node i through node j at time t
$b_i(t)$	Set of offers or demands from node i at time t
t	discrete time value, where $t \in T$
δ_t	Time interval duration
v	number of time values since last update
$x_{ij}(t)$	Decision variable. How much data flows through edge $(i, j) \in E$ at time t

Objective Functions

$$\max \prod_{(i,j)\in P} d_{ij}(t)x_{ij}(t) \tag{2}$$

$$\min \sum_{(i,j)\in P} \sum_{t\in T} x_{ij}(t)\delta_t \tag{3}$$

Constraints

$$\sum_{j\in N} x_{ij}(t) - \sum_{j\in N} x_{ji}(t) + x_{ii}(t) = b_i(t) \qquad \forall(i,j) \in E, i \neq j, t = 0 \tag{4}$$

$$\sum_{j\in N} x_{ij}(t) - \sum_{j\in N} x_{ji}(t) + x_{ii}(t) - x_{ii}(t-1) = b_i(t) \qquad \forall(i,j) \in E, i \neq j, t \geq 1 \tag{5}$$

$$x_{ij}(t) + x_{ji}(t) \leq c_{ij}\delta_t \qquad \forall(i,j) \in E, i \neq j \tag{6}$$

$$x_{ii}(t) \leq c_{ii} \qquad \forall i \in N \tag{7}$$

$$x_{ij}(t) \in \mathbb{Z}_{\geq 0} \qquad \forall(i,j) \in E, p \in P \tag{8}$$

3.2. DISTRIBUTED FORWARDING ALGORITHM

Nodes have to make forwarding decisions with local and limited information. Thus, in our solution, nodes use a distributed forwarding algorithm based on historical and social information. Each node meets its neighbors and gives them an availability probability that will decay over time and will increase when they meet again. Also, when a node meets a neighbor, they share their delivery probabilities to the destination node. Sensor nodes will start with a delivery probability near zero, so that nodes will prefer to store their messages until they contact the Gateway or another node that had been in contact with the Gateway, updating its delivery probability. They will also update their delivery probabilities through time, diminishing them as time goes by without further contacts with the same node. This updating will follow Equation (9). MAPs and the Gateway will start with a delivery probability near 0.9, since they have the links that must be transited on the network. Sensor nodes can use other sensor nodes to deliver their messages to the Gateway if they have a better delivery probability and are available to receive the messages. The Gateway will also update its delivery probability according to its knowledge of the network, giving higher delivery probabilities to MAPs closer to a scheduled or predicted encounter and closer, in time, to deliver the message to the Internet Connected Node.

$$d_{ij}(t+v) = d_{ij}(t)^v \quad (9)$$

Nodes have to save the last contact time, the number of contacts, and calculate the mean time between contacts for every neighboring node. Based on this mean time between contacts, nodes will update the availability probabilities, giving them the maximum value when the contact should be close in time and diminishing them exponentially using Equation (9), wherev is the number of time values since the last update. If the node has multiple neighbors with paths to the destination, then it can make the decision to wait for a certain node to become available or to deliver its message through an already available node, depending on their delivery probabilities and mean time between contacts. We are also using a replication option, where a source node can create up to one copy of a message (there will exist up to two identical messages in the network at any given time) and give a copy to each of the two neighboring nodes with the highest delivery probabilities. Sensor nodes, then, can use other nodes to reach the Gateway. The Gateway can use one of several MAPs that will pass nearby to deliver their messages. The messages will eventually reach their destination, the Central Management and Monitoring Platform (MMS). The Distributed Forwarding Algorithm is shown in pseudo-code in Algorithm 1. Figure 6 shows the tables that the nodes will create to keep their availability and delivery probabilities to their neighbors and messages destinations. The number between parentheses in the availability table is the number of contacts with each node.

4. RESULTS AND DISCUSSION

The results from the Distributed Forwarding Algorithm (DFA) were obtained using the ONE Simulator [24]. Mobility traces used for the simulation were generated by the ONE Simulator and OpenJump [25] over a map depicting the area where the GOLDFISH Colombian tests took place, as can be seen in Figure 7 and Figure 8. The implementation of our algorithm in the ONE Simulator and the files used to run the simulations to compare it with other protocols in this scenario are available at the GitHub repository of this project [26]. Figure 9, Figure 10, Figure 11, Figure 12, Figure 13, Figure 14,Figure 15, Figure 16, Figure 17, Figure 18, Figure 19, Figure 20, Figure 21, Figure 22, Figure 23, Figure 24, Figure 25, Figure 26, Figure 27, Figure 28, Figure 29, Figure 30, Figure 31, Figure 32, Figure 33, Figure 34, Figure 35, Figure

36, Figure 37,Figure 38, Figure 39 and Figure 40 show simulation results.

Algorithm 1 Distributed algorithm for every node in the network.

```
node start (initialization)
broadcast ID
listen for neighbors' IDs
wait for messages()
if message received then
    save neighbors' IDs and contact time in neighbors list
    update mean time between contacts
    update availability and delivery probabilities
    update number of contacts with this neighbor
if neighbor requests delivery probability then
    if final destination = own ID then
        send delivery probability 1
    else if destination known then
        return delivery probability (not through requesting node)
    else if destination unknown then
        do not reply
else if neighbor sends a message then
    if it is for me then
        read message
    else
        store message and send it to next node
else if node has message to send then
    ask for delivery probabilities to destination
    send a copy of the message to the two best delivery probability neighbors
    if no delivery probabilities then
        store message in buffer until a path to destination appears
```

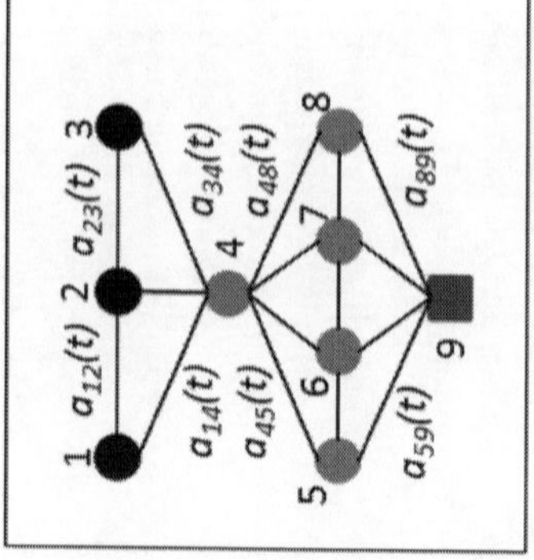

node 1: availability table

node.id	availability probability	last contact time	average time between contacts
2(120)	a_{12}	02:01:20	00:01:00
4(4)	a_{14}	02:00:45	00:30:00

node 1: delivery probabilities table

node.id	delivery probability	next hop	estimated time to contact
9	d_{12}	2	00:01:00
	d_{14}	4	00:29:25

Figure 6. Availability table and delivery probabilities table for each node.

Figure 7. GOLDFISH project map from Google Earth in OpenJump.

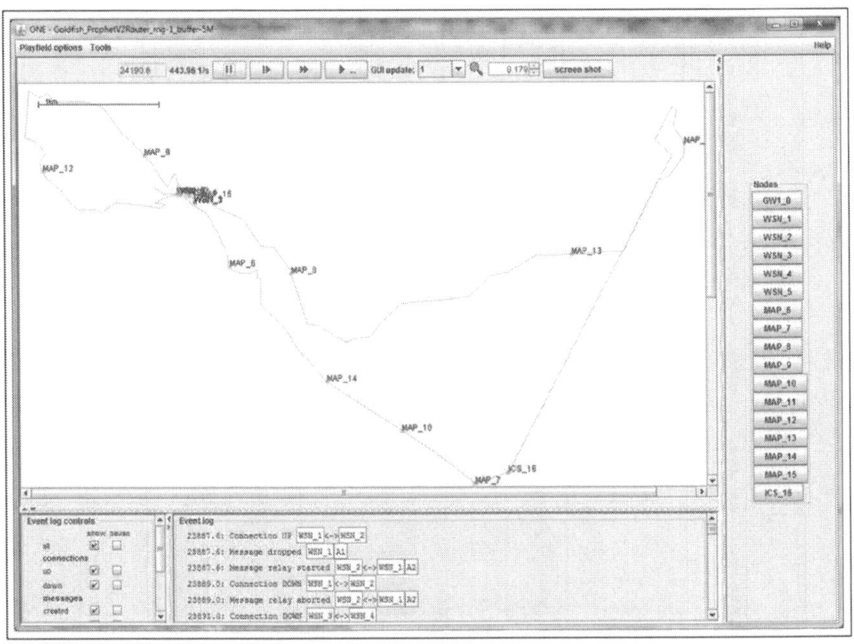

Figure 8. GOLDFISH project map scaled in the ONE Simulator.

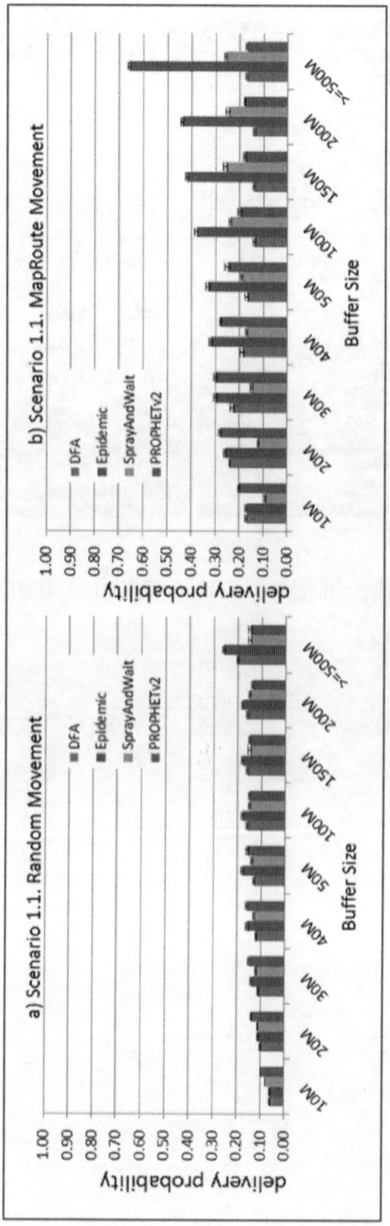

Figure 9. Results for scenario 1.1. Delivery probabilities with different Buffer Sizes, generating 2 MB messages every 1 to 2 min, (**a**) MAPs moving randomly on the path. (**b**) MAPs following the path in only one direction. DFA: Distributed Forwarding Algorithm.

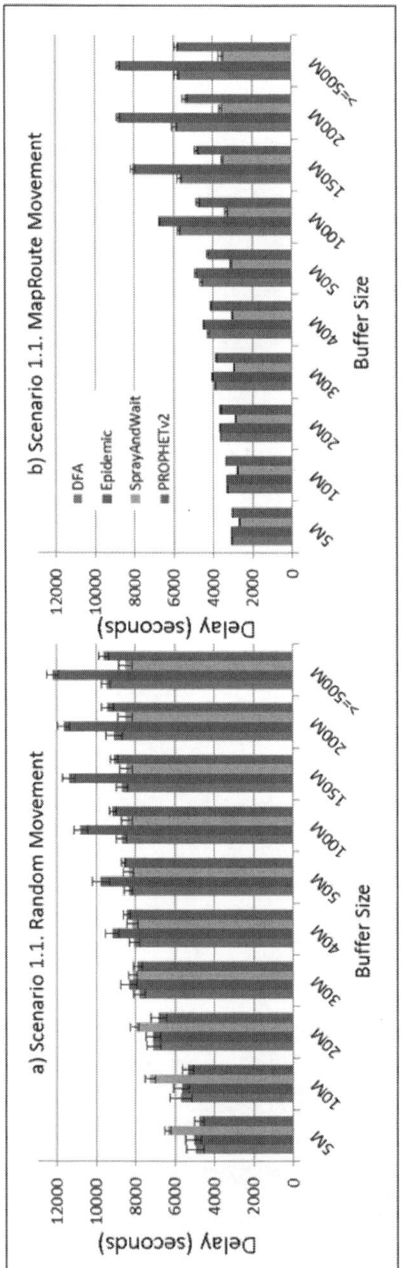

Figure 10. Results for scenario 1.1. Average delay in seconds with different Buffer Sizes, generating 2 MB messages every 1 to 2 min, (**a**) MAPs moving randomly on the path; (**b**) MAPs following the path in only one direction. DFA: Distributed Forwarding Algorithm.

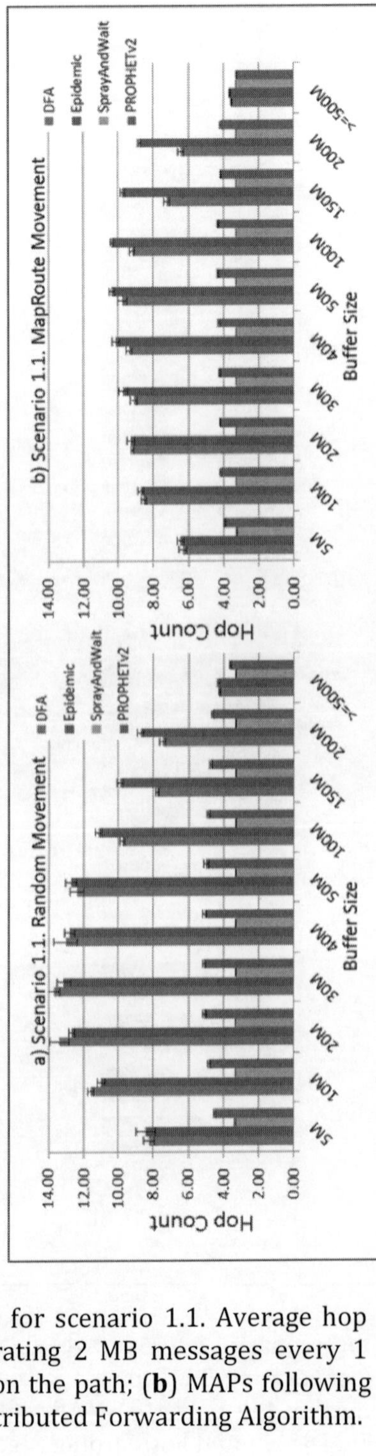

Figure 11. Results for scenario 1.1. Average hop count with different Buffer Sizes, generating 2 MB messages every 1 to 2 min, (**a**) MAPs moving randomly on the path; (**b**) MAPs following the path in only one direction. DFA: Distributed Forwarding Algorithm.

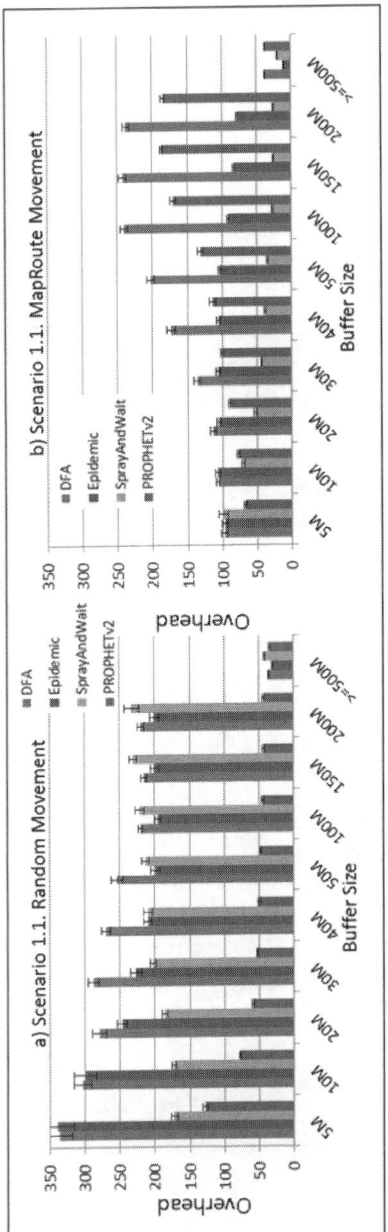

Figure 12. Results for scenario 1.1. Average overhead with different Buffer Sizes, generating 2 MB messages every 1 to 2 min, (**a**) MAPs moving randomly on the path; (**b**) MAPs following the path in only one direction. DFA: Distributed Forwarding Algorithm.

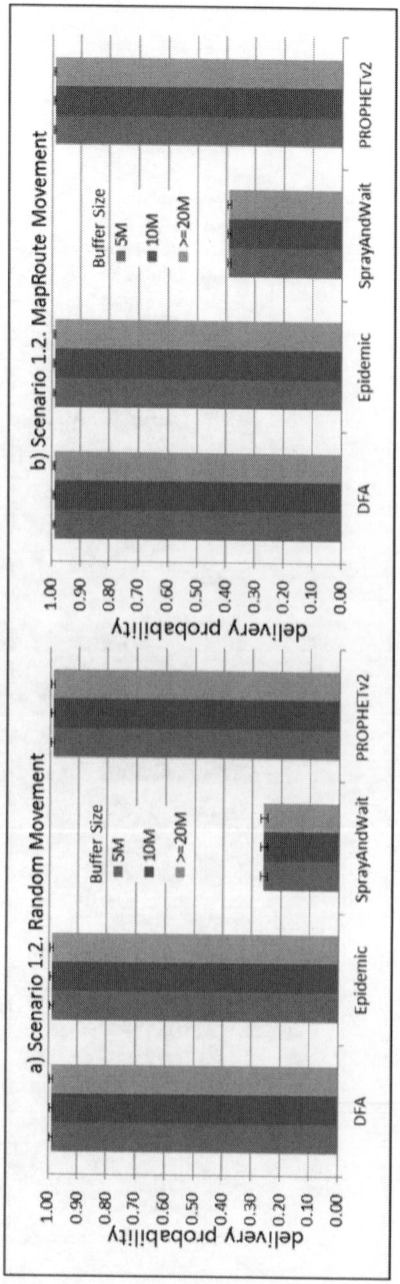

Figure 13. Results for scenario 1.2. Delivery probabilities with different Buffer Sizes, generating 2 kB messages every 1 to 2 min, (**a**) MAPs moving randomly on the path; (**b**) MAPs following the path in only one direction. DFA: Distributed Forwarding Algorithm.

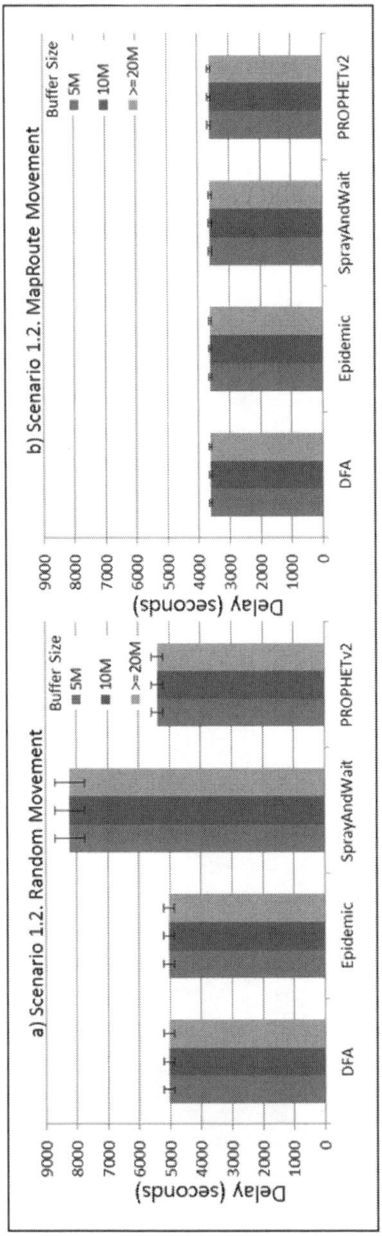

Figure 14. Results for scenario 1.2. Average delay in seconds with different Buffer Sizes, generating 2 kB messages every 1 to 2 min, (**a**) MAPs moving randomly on the path; (**b**) MAPs following the path in only one direction. DFA: Distributed Forwarding Algorithm.

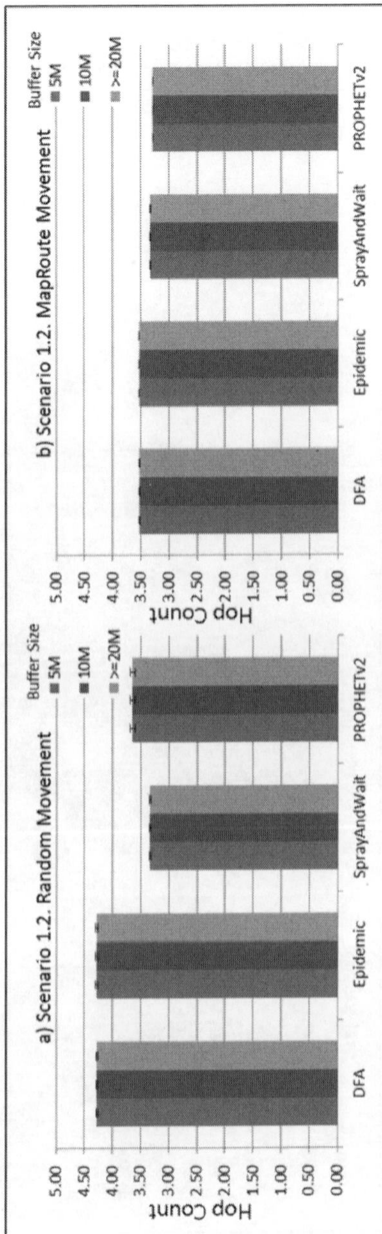

Figure 15. Results for scenario 1.2. Average hop count with different Buffer Sizes, generating 2 kB messages every 1 to 2 min, (**a**) MAPs moving randomly on the path; (**b**) MAPs following the path in only one direction. DFA: Distributed Forwarding Algorithm.

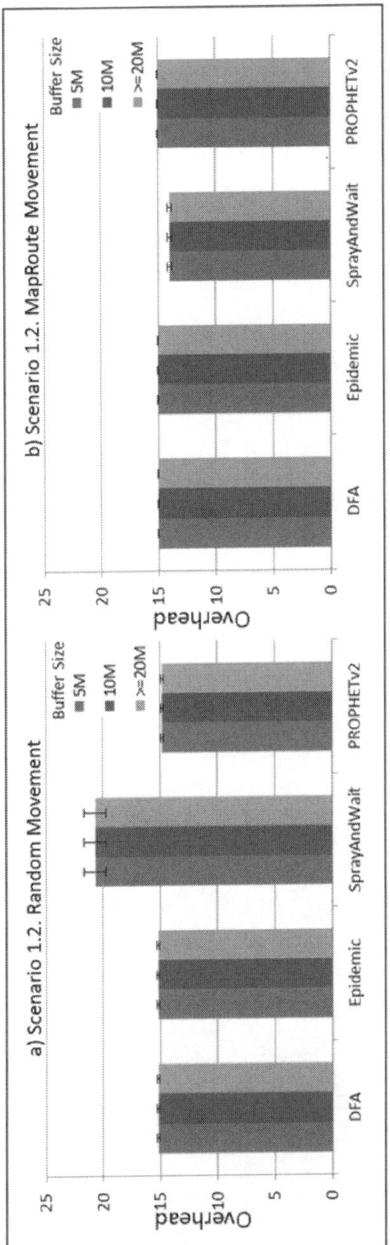

Figure 16. Results for scenario 1.2. Average overhead with different Buffer Sizes, generating 2 kB messages every 1 to 2 min, (**a**) MAPs moving randomly on the path; (**b**) MAPs following the path in only one direction. DFA: Distributed Forwarding Algorithm.

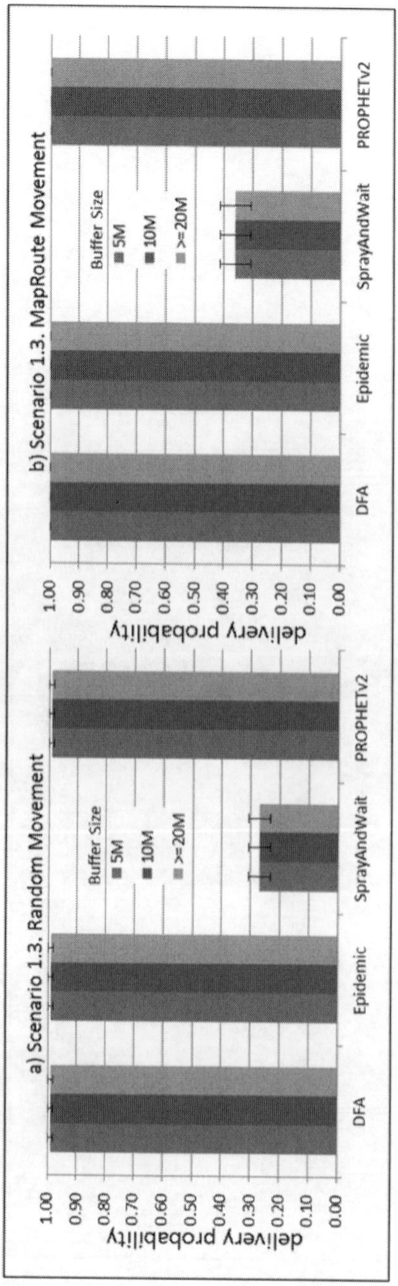

Figure 17. Results for scenario 1.3. Delivery probabilities with different Buffer Sizes, generating 2 kB messages every 1 to 2 h, (**a**) MAPs moving randomly on the path; (**b**) MAPs following the path in only one direction. DFA: Distributed Forwarding Algorithm.

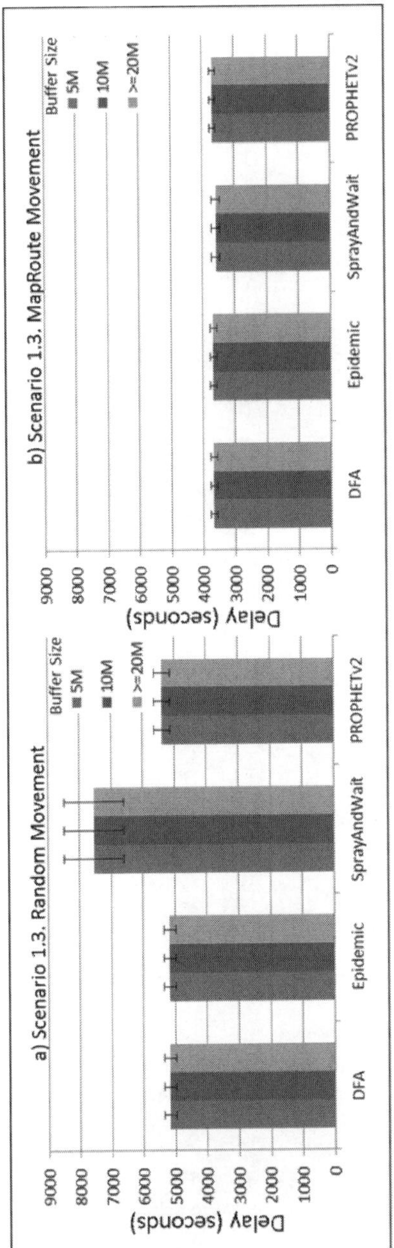

Figure 18. Results for scenario 1.3. Average delay in seconds with different Buffer Sizes, generating 2 kB messages every 1 to 2 h, (**a**) MAPs moving randomly on the path; (**b**) MAPs following the path in only one direction. DFA: Distributed Forwarding Algorithm.

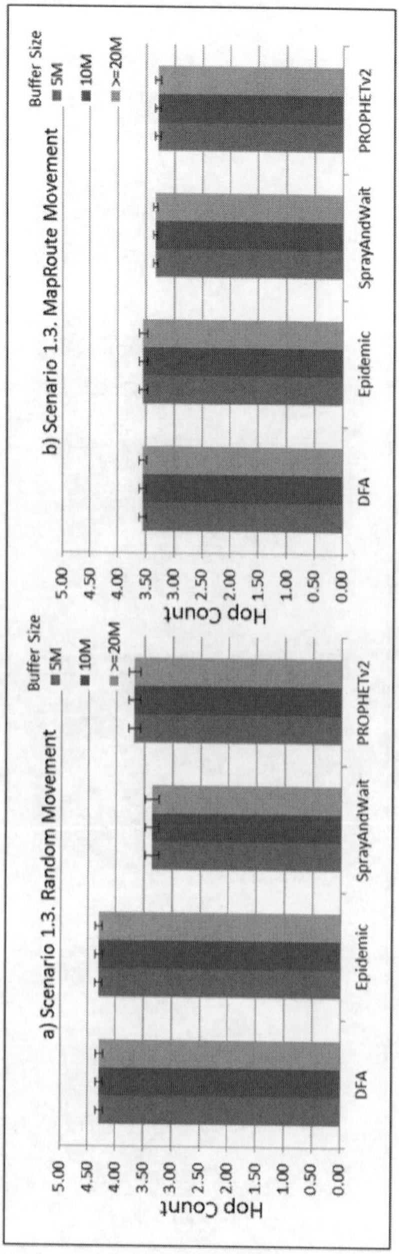

Figure 19. Results for scenario 1.3. Average hop count with different Buffer Sizes, generating 2 kB messages every 1 to 2 h, (**a**) MAPs moving randomly on the path; (**b**) MAPs following the path in only one direction. DFA: Distributed Forwarding Algorithm.

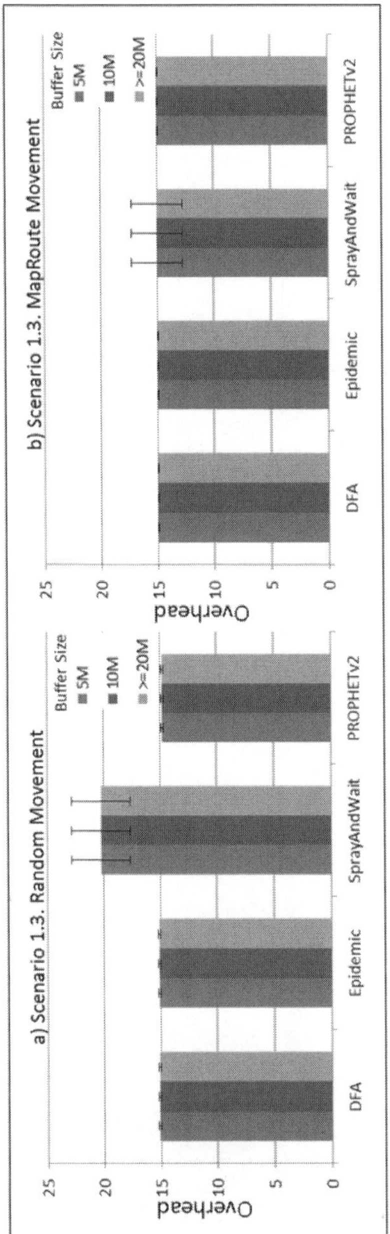

Figure 20. Results for scenario 1.3. Average overhead with different Buffer Sizes, generating 2 kB messages every 1 to 2 h, (**a**) MAPs moving randomly on the path; (**b**) MAPs following the path in only one direction. DFA: Distributed Forwarding Algorithm.

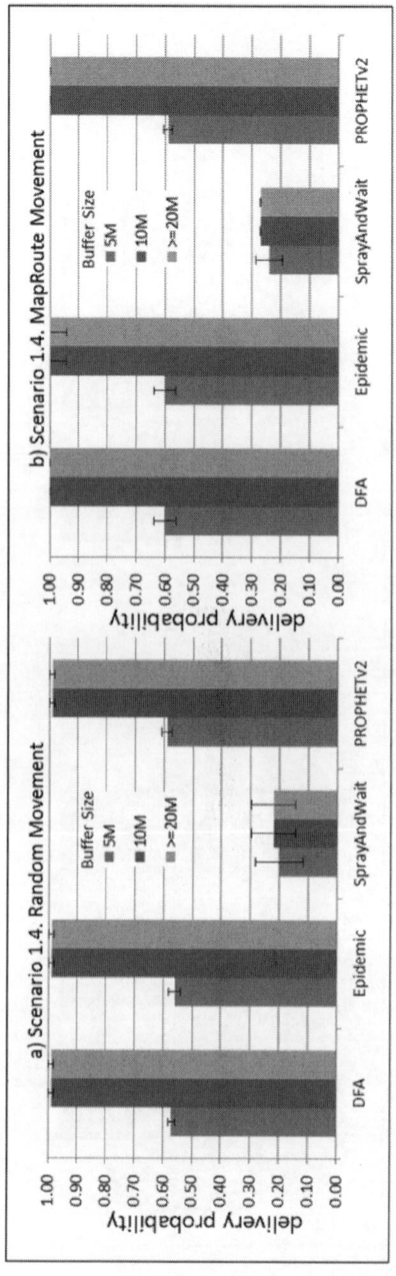

Figure 21. Results for scenario 1.4. Delivery probabilities with different Buffer Sizes, generating 2 MB messages every 1 to 2 h, (**a**) MAPs moving randomly on the path; (**b**) MAPs following the path in only one direction. DFA: Distributed Forwarding Algorithm.

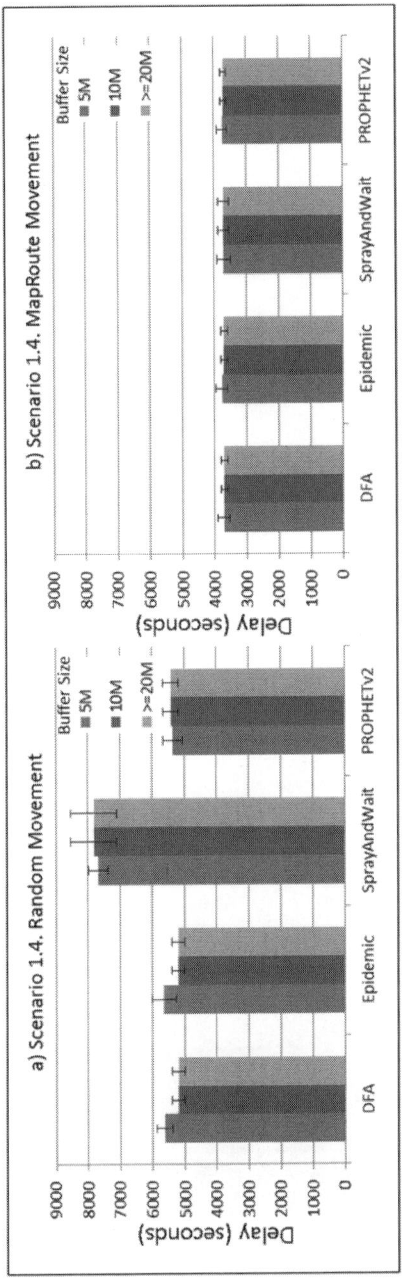

Figure 22. Results for scenario 1.4. Average delay in seconds with different Buffer Sizes, generating 2 MB messages every 1 to 2 h, (**a**) MAPs moving randomly on the path; (**b**) MAPs following the path in only one direction. DFA: Distributed Forwarding Algorithm.

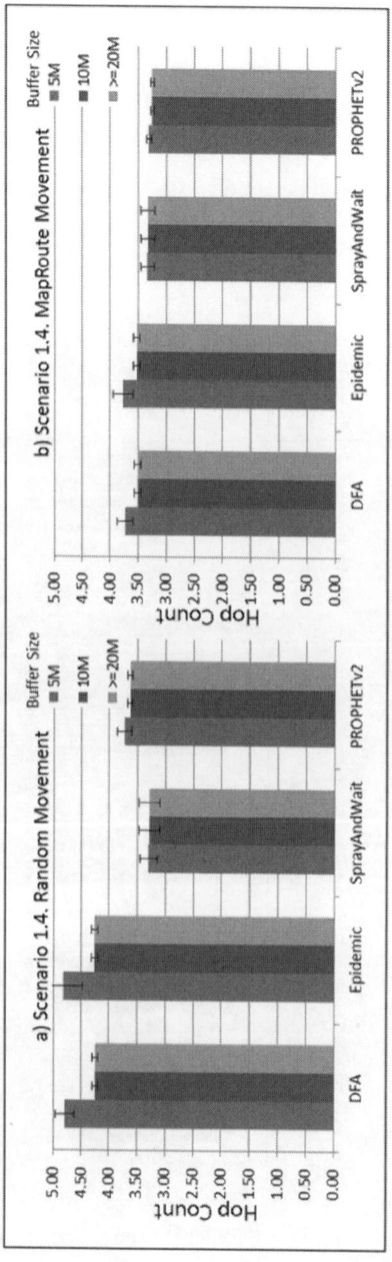

Figure 23. Results for scenario 1.4. Average hop count with different Buffer Sizes, generating 2 MB messages every 1 to 2 h, (**a**) MAPs moving randomly on the path; (**b**) MAPs following the path in only one direction. DFA: Distributed Forwarding Algorithm.

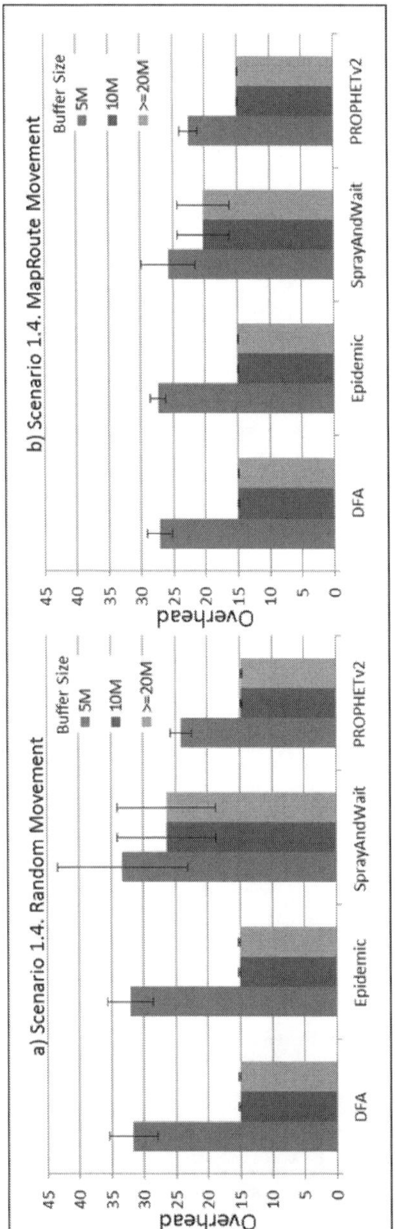

Figure 24. Results for scenario 1.4. Average overhead with different Buffer Sizes, generating 2 MB messages every 1 to 2 h, (**a**) MAPs moving randomly on the path; (**b**) MAPs following the path in only one direction. DFA: Distributed Forwarding Algorithm.

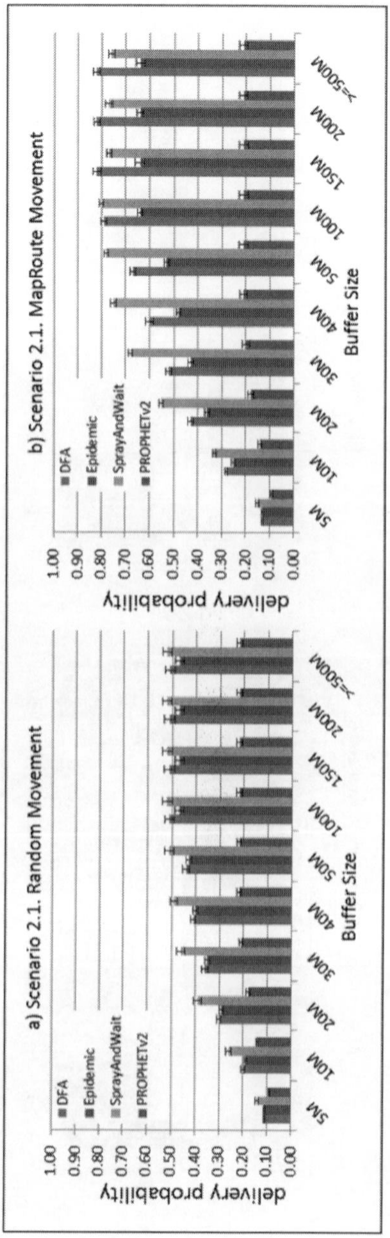

Figure 25. Results for scenario 2.1. Delivery probabilities with different Buffer Sizes, generating 2 MB messages every 1 to 2 min, (**a**) MAPs moving randomly on the path; (**b**) MAPs following the path in only one direction. DFA: Distributed Forwarding Algorithm.

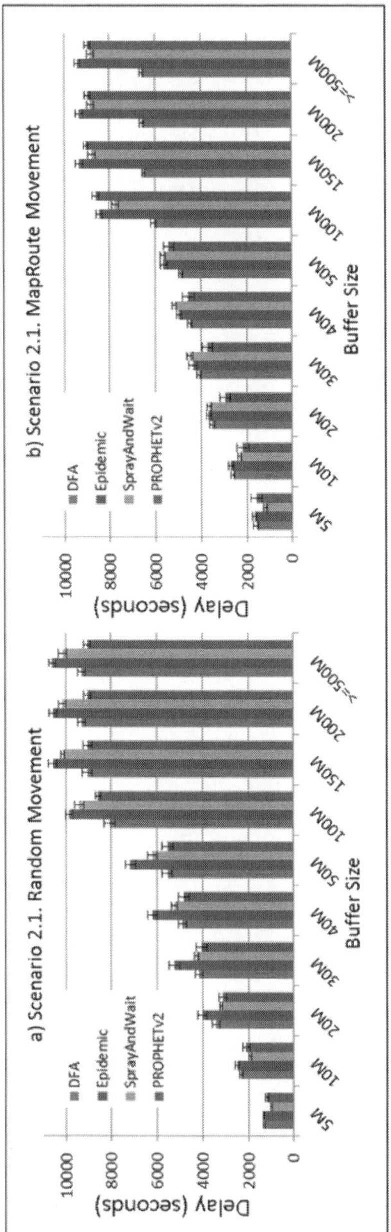

Figure 26. Results for scenario 2.1. Average delay in seconds with different Buffer Sizes, generating 2 MB messages every 1 to 2 min, (**a**) MAPs moving randomly on the path; (**b**) MAPs following the path in only one direction. DFA: Distributed Forwarding Algorithm.

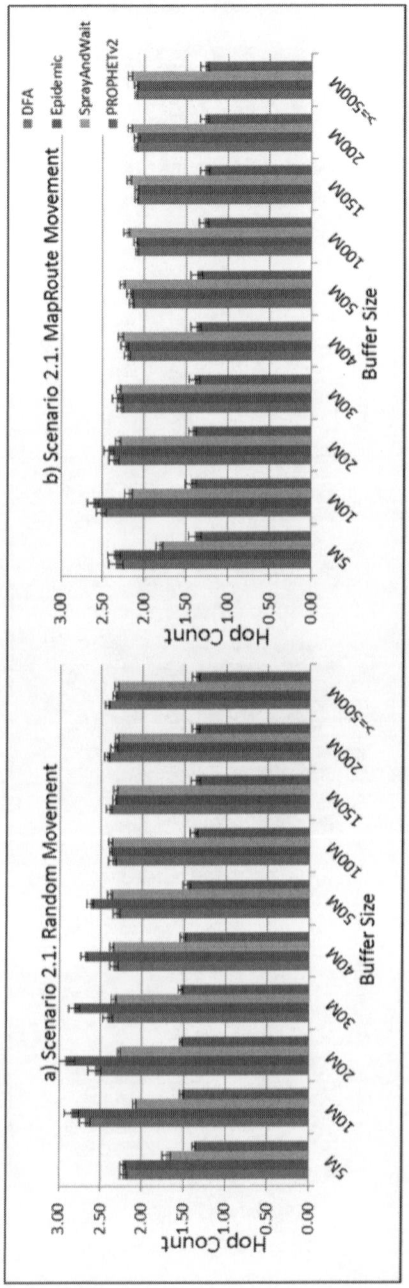

Figure 27. Results for scenario 2.1. Average hop count with different Buffer Sizes, generating 2 MB messages every 1 to 2 min, (**a**) MAPs moving randomly on the path; (**b**) MAPs following the path in only one direction. DFA: Distributed Forwarding Algorithm.

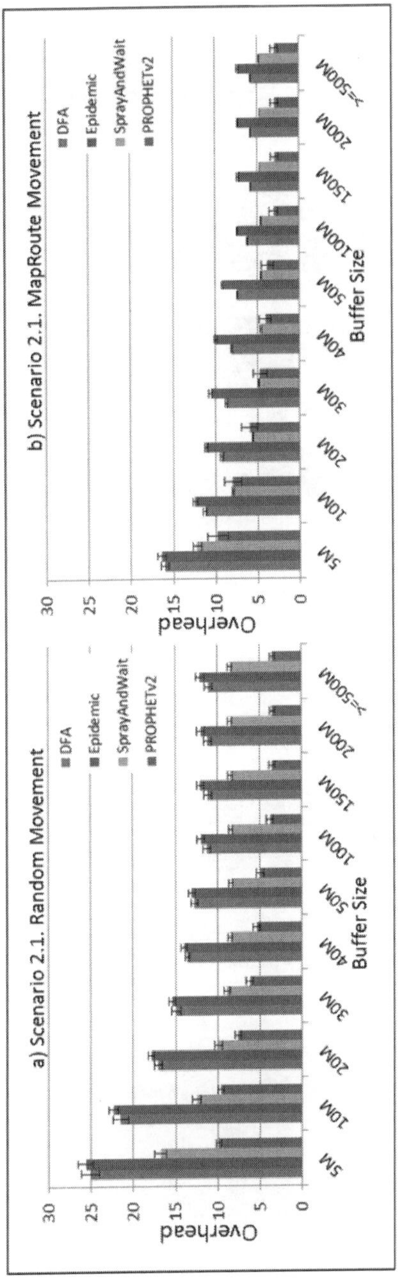

Figure 28. Results for scenario 2.1. Average overhead with different Buffer Sizes, generating 2 MB messages every 1 to 2 min, (**a**) MAPs moving randomly on the path; (**b**) MAPs following the path in only one direction. DFA: Distributed Forwarding Algorithm.

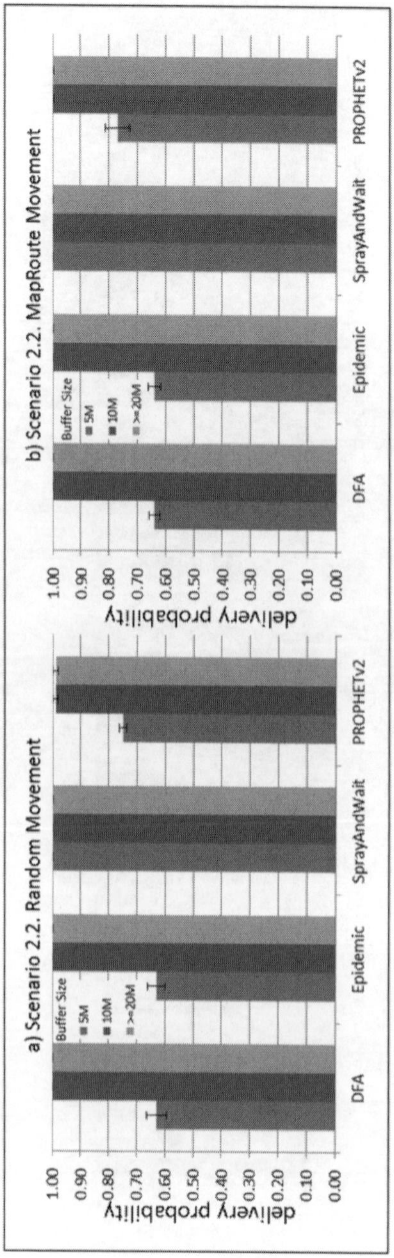

Figure 29. Results for scenario 2.2. Delivery probabilities with different Buffer Sizes, generating 2 kB messages every 1 to 2 min, (**a**) MAPs moving randomly on the path; (**b**) MAPs following the path in only one direction. DFA: Distributed Forwarding Algorithm.

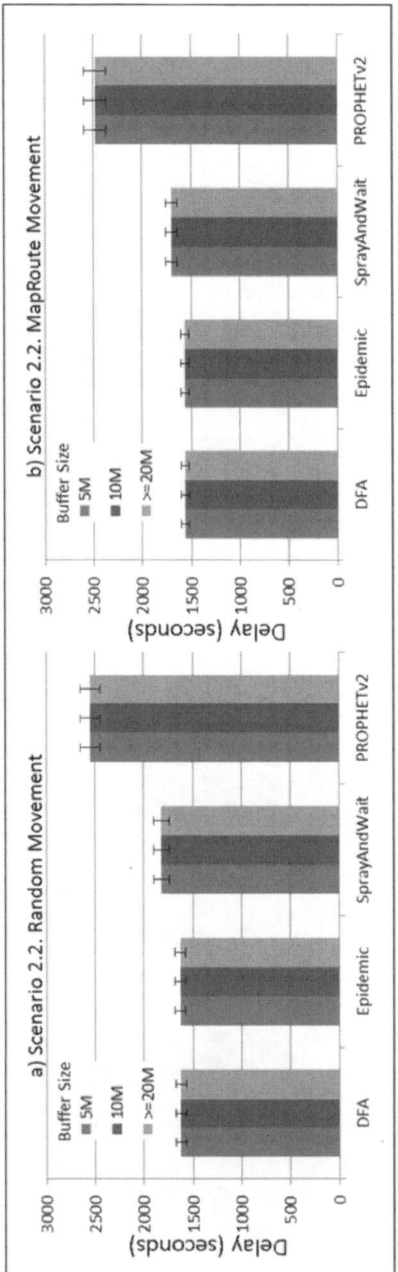

Figure 30. Results for scenario 2.2. Average delay in seconds with different Buffer Sizes, generating 2 kB messages every 1 to 2 min, (**a**) MAPs moving randomly on the path; (**b**) MAPs following the path in only one direction. DFA: Distributed Forwarding Algorithm.

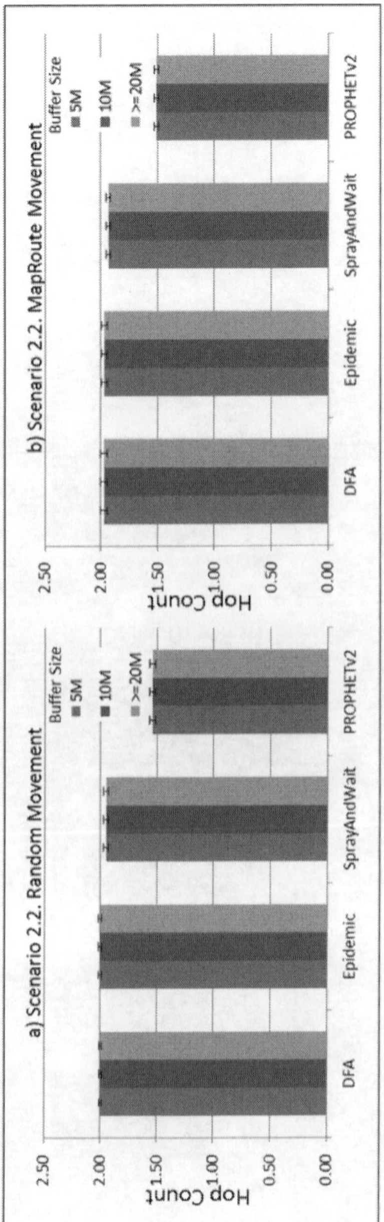

Figure 31. Results for scenario 2.2. Average hop count with different Buffer Sizes, generating 2 kB messages every 1 to 2 min, (**a**) MAPs moving randomly on the path; (**b**) MAPs following the path in only one direction. DFA: Distributed Forwarding Algorithm.

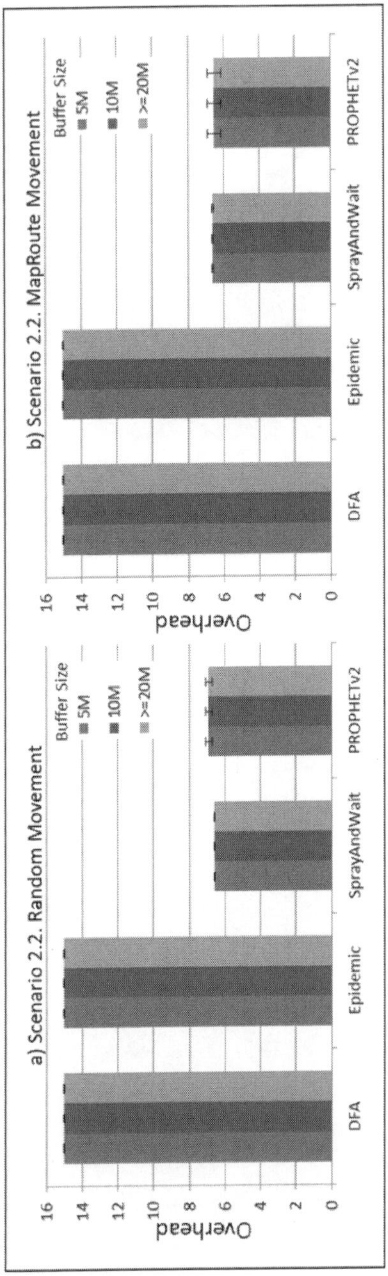

Figure 32. Results for scenario 2.2. Average overhead with different Buffer Sizes, generating 2 kB messages every 1 to 2 min, (**a**) MAPs moving randomly on the path; (**b**) MAPs following the path in only one direction. DFA: Distributed Forwarding Algorithm.

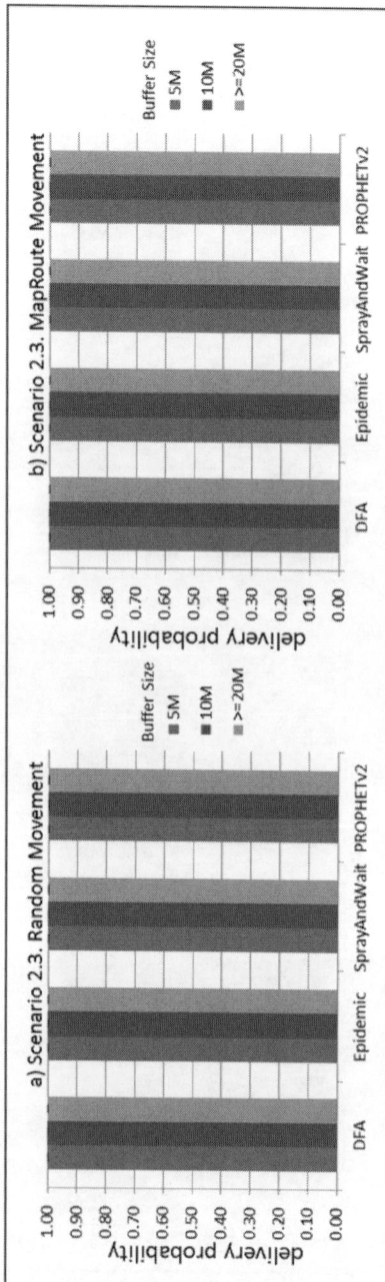

Figure 33. Results for scenario 2.3. Delivery probabilities with different Buffer Sizes, generating 2 kB messages every 1 to 2 h, (**a**) MAPs moving randomly on the path; (**b**) MAPs following the path in only one direction. DFA: Distributed Forwarding Algorithm.

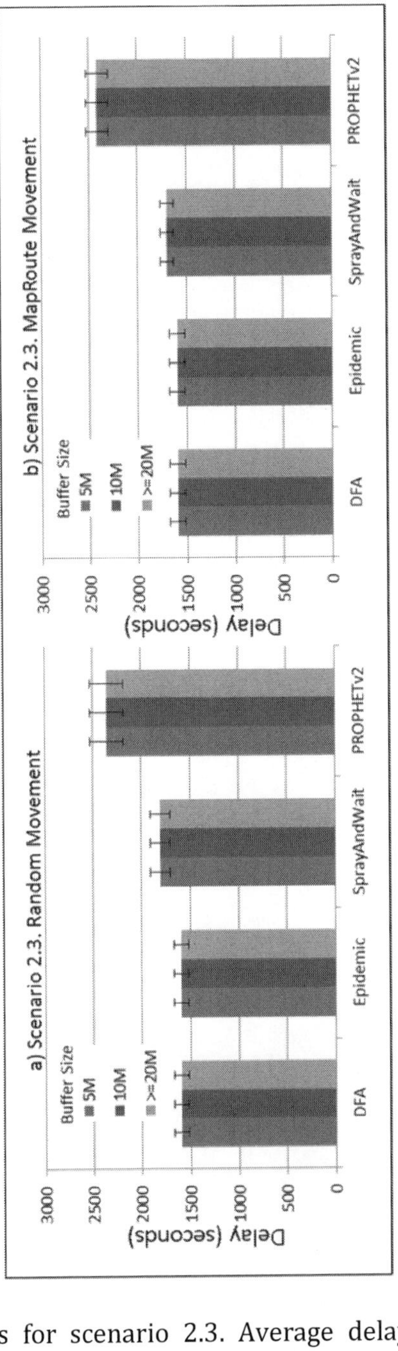

Figure 34. Results for scenario 2.3. Average delay in seconds with different Buffer Sizes, generating 2 kB messages every 1 to 2 h, (**a**) MAPs moving randomly on the path; (**b**) MAPs following the path in only one direction. DFA: Distributed Forwarding Algorithm.

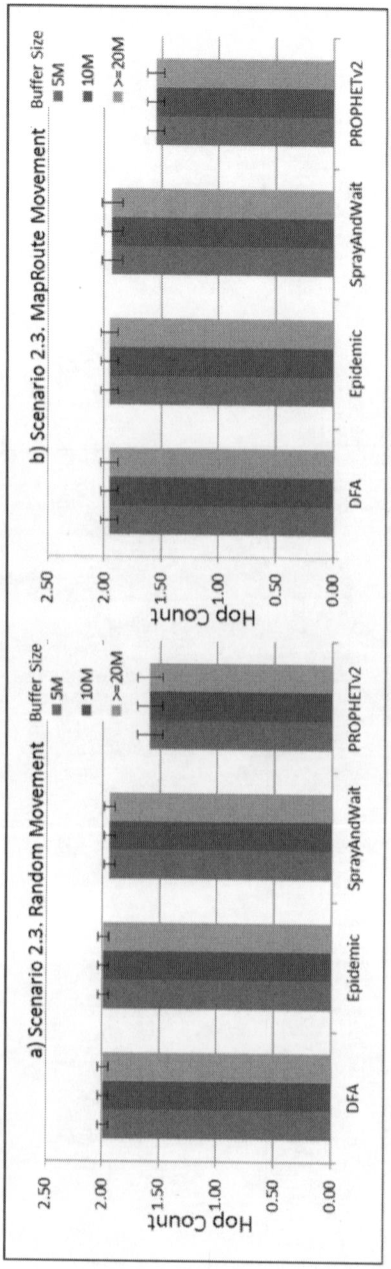

Figure 35. Results for scenario 2.3. Average hop count with different Buffer Sizes, generating 2 kB messages every 1 to 2 h, (**a**) MAPs moving randomly on the path; (**b**) MAPs following the path in only one direction. DFA: Distributed Forwarding Algorithm.

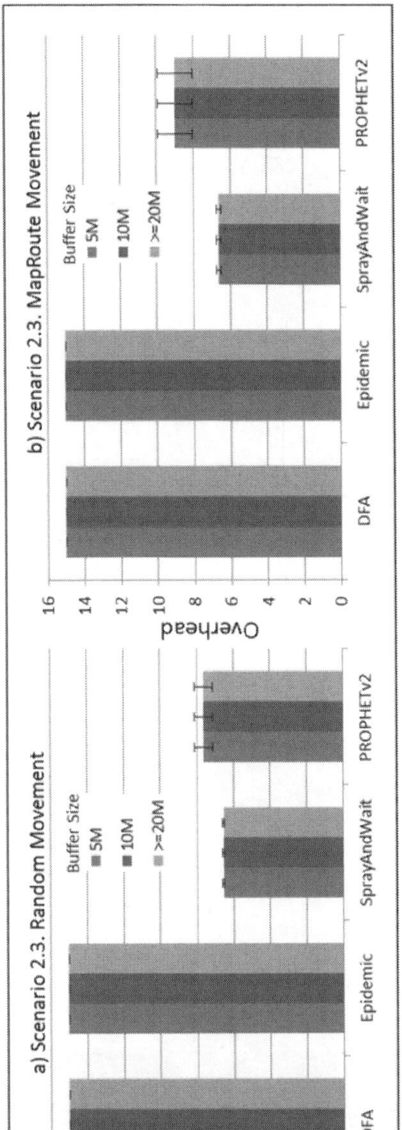

Figure 36. Results for scenario 2.3. Average overhead with different Buffer Sizes, generating 2 kB messages every 1 to 2 h, (**a**) MAPs moving randomly on the path; (**b**) MAPs following the path in only one direction. DFA: Distributed Forwarding Algorithm.

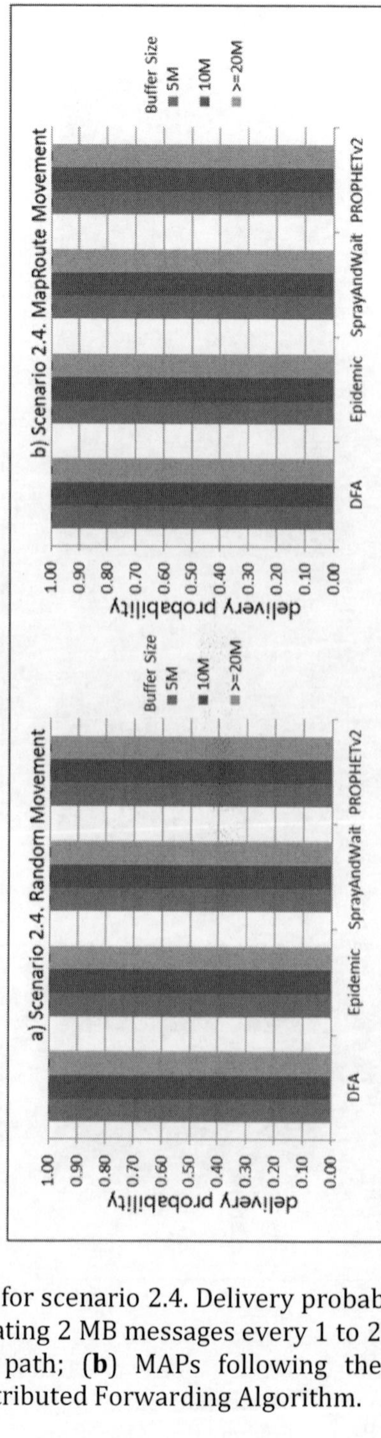

Figure 37. Results for scenario 2.4. Delivery probabilities with different Buffer Sizes, generating 2 MB messages every 1 to 2 h, (**a**) MAPs moving randomly on the path; (**b**) MAPs following the path in only one direction. DFA: Distributed Forwarding Algorithm.

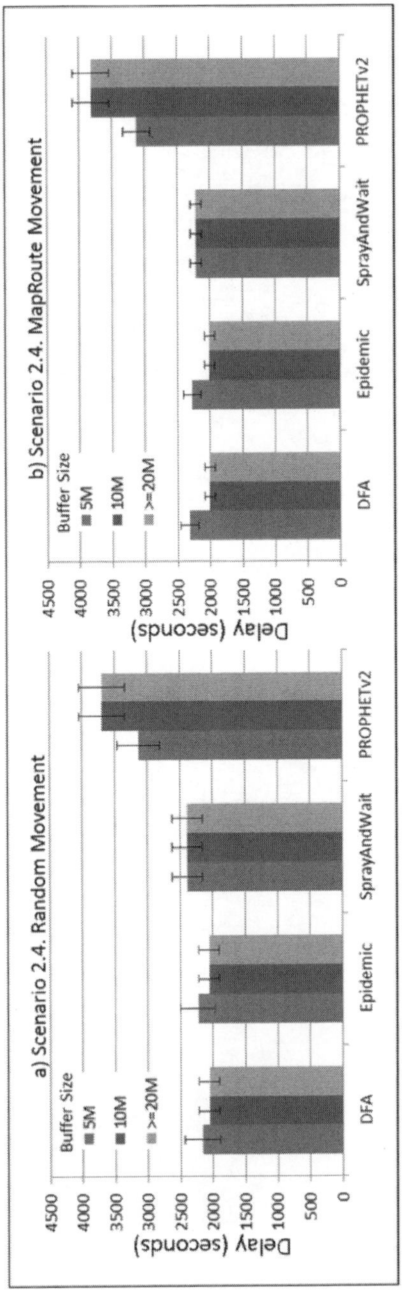

Figure 38. Results for scenario 2.4. Average delay in seconds with different Buffer Sizes, generating 2 MB messages every 1 to 2 h, (**a**) MAPs moving randomly on the path; (**b**) MAPs following the path in only one direction. DFA: Distributed Forwarding Algorithm.

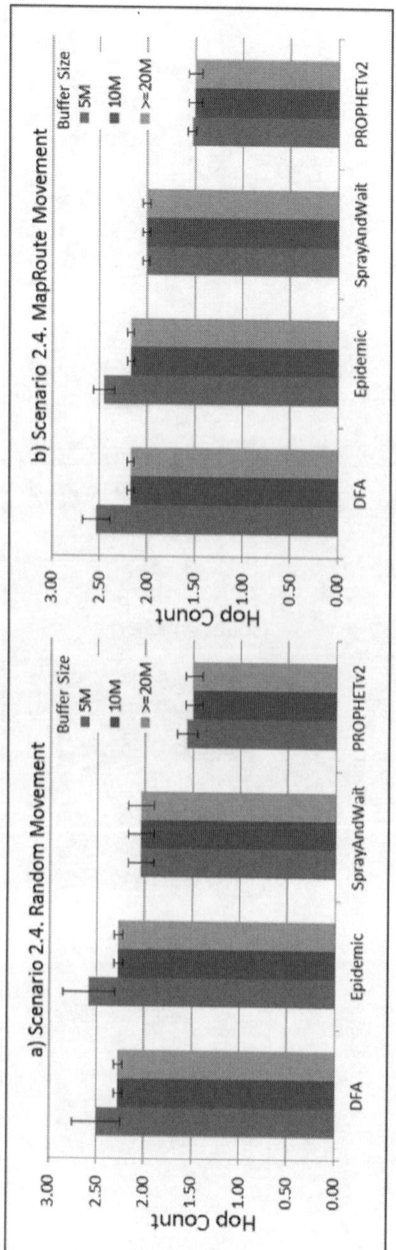

Figure 39. Results for scenario 2.4. Average hop count with different Buffer Sizes, generating 2 MB messages every 1 to 2 h, (**a**) MAPs moving randomly on the path; (**b**) MAPs following the path in only one direction. DFA: Distributed Forwarding Algorithm.

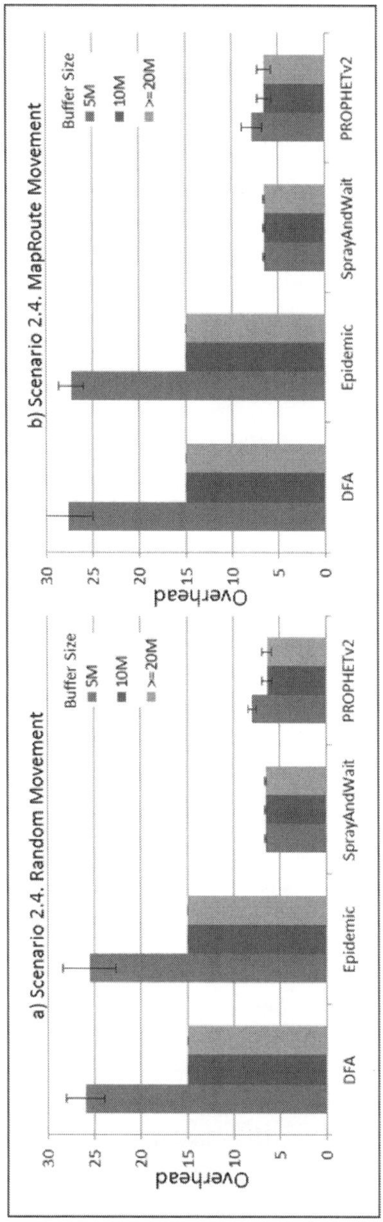

Figure 40. Results for scenario 2.4. Average overhead with different Buffer Sizes, generating 2 MB messages every 1 to 2 h, (**a**) MAPs moving randomly on the path; (**b**) MAPs following the path in only one direction. DFA: Distributed Forwarding Algorithm.

Our Distributed Forwarding Algorithm was tested against three of the most popular and used routing algorithms in DTNs, namely, Epidemic Routing [11], Spray and Wait [12], and PROPHETv2 [9]. Simulation parameters are showed inTable 3. We tested the scenario shown in Figure 7 and Figure 8 with different configurations, as shown in Table 4. In our first scenario, five sensor nodes (in the river) generate information messages that should be sent to the gateway (at the river bank), so it can pass them to the MAPs, and they can eventually reach the ICS. The number of nodes come from the GOLDFISH project, where there are a few nodes on the river communicating with the Gateway at the river shore, and the number of MAPs (10 mobile APs) was chosen based on the idea that there were enough transmission opportunities to take the sensor messages to the ICS and to cover a rural area of approximately 20 km × 16 km where public transportation is scarce. In this scenario, all sensor nodes' messages should go through the Gateway; there is no direct communication from sensor nodes to MAPs. All nodes have wireless communication capabilities resembling IEEE 802.11b with 8 Mbps and a transmission range of 10 m. We also varied the movement model of the MAPs, from random around the path to MapRoute, which means that all MAPs followed the path at different speeds in only one direction. We tested the scenario with the four protocols, with 11 different buffer sizes (from 5 MB to 1 GB), with the two movement models for the MAPs, and with five different random seeds. Each scenario ran 440 times (once per every protocol, every buffer size, every movement model, and every random seed) for one simulated week (60,4800 s) to create the results (Figure 9, Figure 10, Figure 11, Figure 12, Figure 13, Figure 14, Figure 15, Figure 16, Figure 17, Figure 18, Figure 19, Figure 20, Figure 21, Figure 22, Figure 23 and Figure 24). Results are plotted for delivery rate, average delay, average hop count, and average overhead. After this, we ran another scenario setup (Scenario 2) where sensor nodes (message sources) could move randomly around all paths, creating contacts with MAPs and the ICS. In this second setup there is no need for a Gateway, since sensor nodes can communicate directly with the MAPs and the ICS. We also ran the same parameter combinations for this scenario (Figure 25, Figure 26, Figure 27,Figure 28, Figure 29, Figure 30, Figure 31, Figure 32, Figure 33, Figure 34, Figure 35, Figure 36, Figure 37, Figure 38, Figure 39 and Figure 40). The 95% confidence intervals were calculated for every result and they are plotted on the figures.

Table 3. Simulation parameters. ICS: Internet Connected Server. MAP: Mobile Access Point.

Parameter	Definition
Gateway	1 node
Sensor Nodes	5 nodes
ICS	1 node (destination node)
Mobile APs	10 nodes going through the defined path
RandomMvmt	MAPs going through the path in a random fashion
MapRoute	MAPs going through the path following one direction
Buffer Size	Going from 5 MB to 1000 MB. How much information a node can hold
Message Size	Size of each message. It goes from 2 kB to 2 MB
Message Interval	It determines how often messages are generated
SnWCopies	Spray and Wait copies set to 8
DFA	Distributed Forwarding Algorithm (Our proposal)

Table 4. Simulation scenarios. X = 1. Wireless Sensor Network (WSN) nodes in the river. X = 2. WSN nodes all around the map.

Parameter	Subscenario X.1	Subscenario X.2	Subscenario X.3	Subscenario X.4
Message Size	2 MB	2 kB	2 kB	2 MB
Message Interval	1–2 min	1–2 min	1–2 h	1–2 h

Results shown in Figure 9 indicate that, in this particular scenario, with low connectivity and a bottleneck node (the Gateway), DTN-based protocols can experience difficulties to deliver messages when information sources generate large messages (2 MB) at high frequency (less than 2 min between each message). The Gateway, then, becomes a mandatory path for all messages generating at the sensor nodes and seeking to reach the ICS through the MAPs. In the first scenario (Scenario 1.1, Figure 9), sensor nodes are sending 2 MB messages every 1 to 2 min, saturating the buffers and forcing nodes (probably the Gateway) to discard most messages, even with buffers over 500 MB per node. Results in Figure 13 show that when nodes generate light-sized messages (2 kB), DTN nodes can store more messages than in the previous scenario and deliver most of them. Figure 17 and Figure 21 show results for sensor nodes generating differently-sized messages every 1 to 2 h. Here, it can be seen that with this low frequency message generation, DTN-nodes can deliver most messages even with small-sized buffers. The exception in these results are for Spray and Wait protocol, but it can be due to configuration parameters and the bottleneck at the Gateway. The number of copies in Spray and Wait was chosen from simulation results from TheONE simulator, we used the file snw_comparison_settings.txt given in the simulator to compare different values of this parameter. From these simulations we deduced that for copy values over 8, in small sized scenarios, the delivery rate does not change considerably (less than 3 percent). We added this Spray and Wait comparison file to our GitHub project repository [26]. Figure 17 and Figure 21 show that our proposal performs at the same level or even better than Epidemic and PROPHETv2 protocols when the messages are generated every 1 to 2 h and the buffer size per node is equal to or greater than 10 MB. All results show that the delivery probability is slightly better when the MAPs follow the same direction at a constant speed than when they move randomly over the path.

Results for Scenario 2 (Sensor nodes moving randomly over all paths) show that our proposal performs as well as the other routing protocols or even better (in the case where large messages (2 MB) are generated every 1 to 2 min and the nodes have buffers with 100 MB or more). For the other sub-scenarios in Scenario 2, results indicate that all protocols deliver all or almost all messages when sensor nodes can freely move around the map and messages are small or are generated less frequently (1 to 2 h between each new message).

We can also see from Figure 10 that the delay acts as a non-decreasing function of the buffer size for all protocols in the first subscenario of Scenario 1. Spray and Wait shows the minor delay as the buffer increases, followed by PROPHETv2 and our DFA. For all other subscenarios of Scenario 1, all protocols show a very stable behaviour of their delay (Figure 14, Figure 18, and Figure 22). They experience less delay when the MAPs follow one direction in the map.

For Scenario 2, DFA has the best delay performance overall. It can be seen that even in the most demanding subscenario (Subscenario 2.1), DFA increases its delay as the buffer size increases but it does it at a lower rate than the other protocols. For all other subscenarios, the delay stays stable, with DFA and Epidemic having the minor delays, followed by Spray and Wait and PROPHETv2 with the bigger delays.

Regarding HopCount, we can see that for the first subscenario, DFA and Epidemic try to deliver more messages through more hops, PROPHET and Spray and Wait don't change the number of hops when the buffer size changes (Figure 11). For the other subscenarios (Figure 15, Figure 19 and Figure 23), the hop count remains steady, with all protocols very close, but Spray and Wait having the least number of hops, followed by PROPHETv2 and then by DFA and Epidemic.

The hop count for all protocols in all subscenarios of Scenario 2 (Figure 27, Figure 31, Figure 35 and Figure 39) remains very stable. With PROPHETv2 using the minor number of hops and the other protocols having very similar figures between them.

Finally, for Scenario 1, overhead performance is plotted in Figure 12, Figure 16, Figure 20 and Figure 24. In the first subscenario (Figure 12), we can see that our DFA has the worst performance by replicating more messages than it is delivering. PROPHETv2 has the best performance when the MAPs move randomly, and Spray and Wait has the best performance when the MAPs move in only one direction. For all other subscenarios, all protocols have a very similar overhead performance, with Spray and Wait having a bigger overhead in general.

Regarding the overhead performance for Scenario 2, it can be seen from Figure 28, Figure 32, Figure 36 and Figure 40 that Spray and Wait and PROPHETv2 have the best performance for this metric, with similar figures, while DFA and Epidemic have a bigger overhead in general, with almost the same figures between them.

5. CONCLUSIONS

Delay/Disruption Tolerant Networks are a promising technology for rural internet connectivity applications. They can connect people in remote places using non-real-time applications at low costs. Since DTNs can use local transportation infrastructure, they can reach most regions where no networking or even communications infrastructure exists. For our application, we have shown that DTNs are a feasible way to retrieve sensor data from an isolated spot if we can have a vehicle or a set of vehicles that eventually or periodically will reach that spot.

Simulation results show that our Distributed Forwarding Algorithm can deliver most messages (around 99%) using delivery probabilities learned from neighboring nodes, contact history information and replication, and a buffer of 10 MB or more. In the most demanding sub-scenarios, those where 2 MB messages were generated every 1 to 2 min, all protocols performed similarly. However, when we increased the connectivity in the network, giving sensor nodes the opportunity to directly contact the MAPs or the ICS, our proposed algorithm performed better on delivery rate and delay than all other protocols for buffer values over 100 MB.

Due to our proposed DFA implementation of replication and its slow updating of transient delivery probabilities, it uses more hops trying to get the message to its destination and it generates more overhead than Spray and Wait and PROPHETv2, in most cases.

Our algorithm can be used in applications like GOLDFISH where data is taken from a remote site and decision making takes place in another location. We are looking at applications where those decisions do not have to be taken instantly with a single event, but over a lapse of time with historical data—applications like pollution monitoring.

Future work includes researching tuning parameters for the algorithm and a better way to use the contact information to schedule transmission or look for alternative routes. Also, we look forward to performing real tests with the necessary hardware and mobile nodes in a real scenario.

ACKNOWLEDGMENTS

The research leading to these results has received funding from the European Union Seventh Framework Programme (FP7/2007–2013) under grant agreement No. 269985. The author Carlos Velásquez-Villada received funding by the Colombian "Departamento Administrativo de Ciencia, Tecnología e Innovación - Colciencias Conv. 528/2011".

AUTHOR CONTRIBUTIONS

Carlos Velásquez-Villada developed this project as part of his research. Yezid Donoso supervised the research and participated in the revision processes.

CONFLICTS OF INTEREST

The authors declare no conflict of interest.

REFERENCES

1. Rzeszutko, E.; Solano-Donado, F.; Mycek, M. FP7 Project Goldfish—Practical Application of Sensor Networks for Watercourse Pollution Monitoring. In Proceedings of the XXX Krajowe Sympozjum Telekomunikacji i Teleinformatyki (KSTiT'2014), Poznań, Poland, 3–5 September 2014; pp. 1210–1212.

2. Cerf, V.; Burleigh, S.; Hooke, A.; Torgerson, L.; Durst, R.; Scott, K.; Fall, K.; Weiss, H. Delay-tolerant networking architecture. *IETF Trust* **2007**, *4838*, 1–35.

3. Berners-Lee, T.; Fielding, R.; Masinter, L. Uniform Resource Identifier (URI): Generic Syntax. *Internet Soc.* **2005**, *3986*, 1–60.

4. Scott, K. L.; Burleigh, S. Bundle Protocol Specification. Available online: https://tools.ietf.org/html/rfc5050 (accessed on 13 March 2016).

5. Project Loon. Google Inc. Available online: https://www.google.com/loon/ (accessed on 18 February 2016).

6. Internet.org. Facebook Inc. Available online: https://www.internet.org/projects (accessed on 18 February 2016).

7. Pentland, A.; Fletcher, R.; Hasson, A. Daknet: Rethinking connectivity in developing nations. *Computer* **2004**, *37*, 78–83.

8. Watkins, J.; Tacchi, J.; Kiran, M.S. The Role of Intermediaries in the Development of Asynchronous Rural Access. In Universal Access in Human-Computer Interaction. Applications and Services, Proceedings of the 5th International Conference, UAHCI 2009, Held as Part of HCI International 2009, San Diego, CA, USA, 19–24 July 2009; pp. 451–459.

9. Lindgren, A.; Doria, A.; Davies, E.; Grasic, S. Probabilistic Routing Protocol for Intermittently Connected Networks. Available online: https://tools.ietf.org/html/rfc6693 (accessed on 13 March 2016).

10. Gomez, C.; Paradells, J. Urban Automation Networks: Current and Emerging Solutions for Sensed Data Collection and Actuation in Smart Cities. *Sensors* **2015**, *15*, 22874–22898.

11. Vahdat, A.; Becker, D. *Epidemic Routing for Partially Connected Ad Hoc Networks*; Technical Report CS-200006; Duke University: Durham, NC, USA, 2000.

12. Spyropoulos, T.; Psounis, K.; Raghavendra, C.S. Spray and Wait: An Efficient Routing Scheme for Intermittently Connected Mobile Networks. In Proceedings of the 2005 ACM SIGCOMM workshop on Delay-tolerant networking, Philadelphia, PA, USA, 22–26 August 2005.

13. Hui, P.; Crowcroft, J.; Yoneki, E. BUBBLE Rap: Social-based Forwarding in Delay Tolerant Networks. In Proceedings of the ACM MobiHoc'08, Hong Kong, China, 26–30 May 2008.

14. Zhu, J.; Cao, J.; Liu, M.; Zheng, Y.; Gong, H.; Chen, G. A Mobility Prediction-Based Adaptive Data Gathering Protocol for Delay Tolerant Mobile Sensor Network. In Proceedings of the Global Telecommunications Conference, 2008, (IEEE GLOBECOM 2008), New Orleans, LA, USA, 30 November–4 December 2008.

15. Daly, E.M.; Haahr, M. Social network analysis for routing in disconnected delay-tolerant manets. In Proceedings of the ACM MobiHoc'07, Montreal, QC, Canada, 9–14 September 2007.

16. Mtibaa, A.; May, M.; Diot, C.; Ammar, M. Peoplerank: Social opportunistic forwarding. In Proceedings of the IEEE INFOCOM'10, San Diego, CA, USA, 15–19 March 2010.

17. Balasubramanian, A.; Levine, B.; Venkataramani, A. Replication routing in DTNs: A resource allocation approach.*IEEE ACM Trans. Netw.* **2010**, *18*, 596–609.

18. Tang, L.; Zheng, Q.; Liu, J.; Hong, X. Smart: A selective controlled flooding routing for delay tolerant networks. In Proceedings of the Fourth International Conference on Broadband Communications, Networks and Systems, 2007, BROADNETS 2007, Raleigh, NC, USA, 10–14 September 2007.

19. Spyropoulos, T.; Turletti, T.; Obraczka, K. Routing in delay-tolerant networks comprising heterogeneous node populations. *IEEE Trans. Mob. Comput.* **2009**, *8*, 1132–1147.

20. Cao, Y.; Sun, Z. Routing in Delay/Disruption Tolerant Networks: A Taxonomy, Survey and Challenges. *IEEE Commun. Surv. Tutor.* **2013**, *15*, 654–677.

21. Khabbaz, M.J.; Assi, C.M.; Fawaz, W.F. Disruption-Tolerant Networking: A Comprehensive Survey on Recent Developments and Persisting Challenges. *IEEE Commun. Surv. Tutor.* **2012**, *14*, 607–640.

22. Psaras, I.; Wood, L.; Tafazolli, R. *Delay-Disruption-Tolerant Networking: State of the Art and Future Challenges*; Technical Report; University of Surrey: Guildford, UK, 2010.

23. Velásquez-Villada, C.; Solano, F.; Donoso, Y. Routing Optimization for Delay Tolerant Networks in Rural Applications Using a Distributed Algorithm. *Int. J. Comput. Commun. Control* **2015**, *10*, 100–111.

24. Keränen, A.; Ott, J.; Kärkkäinen, T. The ONE Simulator for DTN Protocol Evaluation. In Proceedings of the 2nd International Conference on Simulation Tools and Techniques (ICST), Rome, Italy, 2–6 March 2009.

25. OpenJump GIS (2016). Available online: http://www.openjump.org/ (accessed on 18 February 2016).

26. Velásquez-Villada, C. DFA Project Repository (2016). Available online: https://github.com/cevelasquezv/DFA_GOLDFISH.git (accessed on 14 March 2016).

7

ANALYTICAL EVALUATION OF SPACE-TIME M-QAM SIGNALING AND TIME-DELAY ESTIMATION IN MULTIPATH TERRESTRIAL COMMUNICATIONS

Ardavan Rahimian

School of Electronic, Electrical and Computer Engineering,
University of Birmingham, Birmingham, UK.

ABSTRACT

This paper presents novel information-theoretic PHY layer performance improvement for combating the system fading effects in wireless terrestrial communication applications. Robust space-time system strategies in cooperation with the high performance unpunctured serially concatenated codes have been jointly employed. The wireless terrestrial system also uses M-ary quadrature amplitude modulation (QAM) signaling to develop a well-established system modeling for multiple-input single-output (MISO) and multiple-input multiple-output (MIMO) analysis. This research also undertakes the analytical evaluation of time delay estimation in the presence of additive white Gaussian noise (AWGN) in multipath areas. The objective of this investigation is to develop mathematically-oriented estimations intended for a suitable employment in terrestrial communications. This model considers the problem of overlapping the terrestrial signals arrival-times estimation from a noisy received waveform in order to develop a system model for terrestrial TDE.

KEYWORDS

MIMO; STBC; TDE; Terrestrial; Turbo Code

1. INTRODUCTION

The increasing demands for improving physical (PHY) layer system performance of wireless systems constitute a thriving global industry. Hence, in order to fulfill PHY layer QoS requirements in modern terrestrial systems, the appropriate combination of wireless technologies for constituting an engineering trade-off in terms of the overall system's bit error rate (BER), and the estimated complexities should be considered. After the introduction of the Alamouti dual-antenna space-time codes [1] which appeal in terms of its performance and complexity, the proposed scheme motivated Tarokh to generalize the Alamouti's scheme to an arbitrary number of antennas [2, 3], leading to the concept of the orthogonal space-time block codes (STBCs) [4]. STBCs have then evolved at an exceptional rate and reached a state of maturity in theory after their breakthrough discovery; they quickly found their way into the practical applications and standards for enabling outstanding reliable technologies such as: LTE, WLAN, WMN, WiMAX, and RFID [5-8]. In order to satisfy the reliability requirements in the systems, highperformance forward error correction (FEC) codes are required in addition to transmit/receive spatial diversity. Turbo codes [9] brought about a wealth of theoretical knowledge and practical solutions for achieving coding gains with feasible decoding complexities. They have been adopted in various wireless standards due to nearcapacity performance, including the ITU-proposed 3G standards [10], ETSI-DVB system [11], UMTS [12], and LTE [13]. In this research, performance analysis of the dual antenna space-time codes based on the Alamouti along with the Tarokh's orthogonal STBCs has been evaluated in cooperation with the high-performance turbo coding scheme. The performance evaluations have been carried out over multiple-input single-output (MISO) and multiple-input multiple-output (MIMO) systems. These systems are operating in the statistical wireless terrestrial communications propagation environment. Hence, terrestrial reliable wireless system deployment is necessary in order to provide highly accurate and efficient services, and is also of crucial importance for modern wireless technologies. The enormous growth in terrestrial communications and with the increasing demand for the highperformance services raise the need for systems analysis and planning for the intended environments to evaluate performance and its operation estimation. The problem of wireless time-delay estimation (TDE) is of considerable research interest in the wireless communications, which is related to the multipath in specified environments. Multipath causes a single transmitted signal originnating from a source to arrive at a receiver via two or more different paths along with any combination of several distinct unequal length paths. These extra paths at a receiver can be synchronized with the mean scaling of the most dominant path and all the others are treated as the noise; then possible equalization strategies can be employed to improve the receiver system performance [14,15]. It is assumed that the transmitted radio signal and the number of paths in the multipath environment are known. The received radio signal is not just the transmitted one, but several delayed and amplitude-weighted versions of the transmitted signal [16].

2. TERRESTRIAL COMMUNICATION TDE THEORY

The problem of multipath time-delay estimation (MTDE) in this research has been characterized using (1) as:

$$\underrightarrow{yields}\ r(t) = \sum_{1}^{M} a_k s(t - \tau_k) + \tilde{\omega}(t);\ 0 \le t \le T. \tag{1}$$

where s(t) is the transmitted radio signal (system pulse); a_k is the amplitude (attenuation value) for path k; t_k is the time-delay for path k; M is the number of different paths; and $\tilde{\omega}(t)$ is the white Gaussian noise (AWGN) in the multipath environment. The correlation receiver is used to obtain the time-delay estimates. If the TDEs are estimated in a multipath environment where paths are spaced closer than a correlation time of a transmitted radio signal, the receiver cannot separate the closely spaced delays. When the terrestrial radio structure is utilized for the estimation, fading effects can be alleviated and significant wireless system performance improvement can be achieved [17].

3. TERRESTRIAL COMMUNICATION SYSTEM MODEL

Figure 1 shows block diagram of the STBC coded wireless MIMO system of interest, employing QAM as the modulation scheme, and turbo codes as a state-of-the-art FEC. The information bit sequence is first fed into a turbo encoder and then modulated in a multilevel QAM modulator where the coded system bit stream is first blocked into a symbol of Log_2^M data bit i.e. four data bits in this case; it should be considered as an appropriate operating point in system power versus spectral efficiency trade-off, since 16-QAM have moderate theoreticcal bandwidth efficiency in M-QAM signaling scheme [4 (bit/sec)/Hz] that is 1 (bit/sec)/Hz for 2-QAM (i.e. BPSK), and 8 (bit/sec)/Hz for 256-QAM, while yields moderate power efficiency compared with other M-QAM schemes according to the theoretical BER bounds for uncoded M-QAM over the additive white Gaussian noise (AWGN) channel, for resulting in the BER of 10^{-6}, and 16-QAM operating SNR is 14.5 dB, while that is 10.5 dB and 23.5 dB for M = 2 and M = 256 schemes respectively. Hence, it has been introduced in various standards such as DVBTerrestrial, HSDPA protocol, and IEEE 802.11a PHY for modulating OFDM subcarriers. The modulated symbols are mapped using space-time block encoder which has advantages of both the spatial diversity provided by multiple antennas, and the temporal diversity available with time-varying fading [18], without channel state information (CSI) [19]. At the receiver, the combiner combines received signals which are then sent to the maximum likelihood (ML) detector. The detected symbols are then demodulated by the multilevel QAM demodulator and are sent to the iterative turbo decoder for recovering the transmitted information bits in the terrestrial radio system transmissions.

4. TERRESTRIAL COMMUNICATION FEC CODING

Turbo codes have attracted a great interest after their advent [9] since they were the first class of error correcting codes that could perform within a fraction of a decibel to capacity limit [20]. The classic turbo coding is a channel coding technique based on the idea of producing long and random code which employs an iterative decoding algorithm to result in BER performance near capacity limit. The encoder of classic turbo codes is constructed by parallel concatenating two recursive systematic convolutional (RSC) codes through an interleaver. They have been termed as parallel concatenated codes (PCCs) as well. The decoder consists of an interleaver linked soft-input soft-output constituent decoders for corresponding RSC codes. Owing to the presence of the system's interleaver, optimum ML decoding algorithms for turbo codes are very complex. However, iterative decoding of codes in which soft-information is exchanged between constituent decoders is quite feasible and leads to near capacity performance. This information could be regarded as a type of diversity that can refine the constituent decoder's outputs in iterations and in the TDE analysis [21]. An enhanced iterative turbo decoder structure using a modified version of the conventional maximum a posteriori (MAP) algorithm [22] was invoked to decode the introduced PCCs. For preserving the performance of the original MAP decoding with the relatively lower complexities, practical reduced complexity decoding approaches were then proposed. These system decoding algorithms introduced include Max Log-MAP [23], Log-MAP [24], and Erfanian scheme [25]. After achieving the system error correction characteristics close to the Shannon limit at a BER of 10^{-5} (at E_b/N_0 = 0.7 dB with code rate of 1/2) based on PCCs, a vast body of efforts are dedicated to develop the codes.

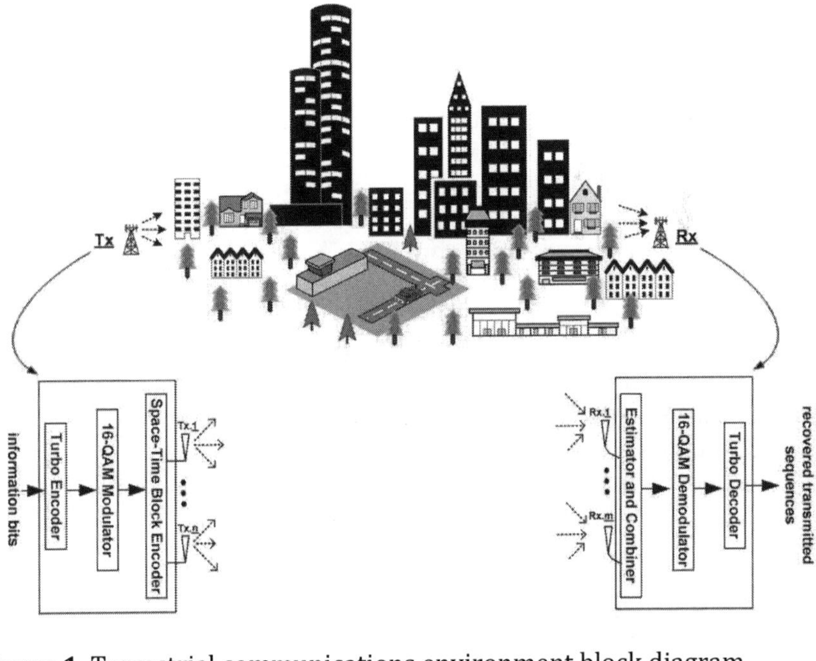

Figure 1. Terrestrial communications environment block diagram.

Unfortunately, in many turbo coded wireless systems, it is reported that for BERs lower than 10^{-5}, the increase in transmit power just marginally reduces the BER values. As a strong technique to sharpen the system BER curve in error-floor region, turbo codes based on serial concatenation of convolutional codes were introduced which reported excellent results from this scheme in moderate to high SNRs [26]. Serial turbo codes which are referred to as serially concatenated codes (SCCs) behave significantly better and do not exhibit the typical error-floor behavior of the PCCs. In cases where relatively smaller interleaver size is employed, PCC's error-floor starts at a much higher BER value of 10^{-3}, while the SCCs have shown to exhibit sharp curve for BERs lower than 10^{-3} [27]. The SCC's encoder consists of the cascade of an outer convolutional code, an interleaver, and also an inner convolutional code whose system input words are the permuted version of outer codewords. The turbo system decoding algorithm in both the SCCs and PCCs are the same. However, it should be noted that the primary difference is that while in the structure of PCCs both the constituent decoders make use of direct output from the demodulator, in SCC's decoder only the inner decoder makes use of this information directly, and the outer one receives the relative extrinsic system information from the deinterleaved codewords of inner decoder's outputs. In the context of designing SCCs, the following parameters have key roles in the turbo coded BER performance: constituent code; interleaver design; and decoding algorithm [28]. A practical performance parameter combination in the design of constituent codes and iterative decoder is utilized for yielding promising tradeoff in terms of BER and complexity. SCCs have been also employed based on quadratic interleavers proposed in [29] which have not only shown to have better BER performance in both waterfall and error-floor regions compared with the classic interleavers [30], but also have straightforward system implementation [31]. In [30], extra coding gain of 2.5 - 3 dB for yielding the BER of 10^{-5} has been reported, when quadratic interleaver is used in SCCs for both the SISO and MISO systems over Rice multipath fading channel; compared with the wireless communication system using matrix-based block interleavers with large and comparable system depth and span.

5. TERRESTRIAL SPACE-TIME CODING

For suppressing multipath effects, diversity techniques like time, frequency, and spatial diversity could be employed. Recent multiple antenna systems based on the space-time codes in which devices are equipped with the multi-transmit and receive antennas, have received considerable attention for research, not only from the theoretical aspects, but also in development and implementation for wireless systems. They are expected to play a major role in futuristic wireless communication systems due to their high diversity order and low decoding complexities, which lead to significantly improve reliability and dramatically increase capacity of wireless systems links [32-35]. Space-time encoding is performed according to the number of the transmit antennas, and the code rate. For rate one of the twin-antenna space-time codes, denoting the signal transmitted from the first and second antennas by s_1 and s_2, the encoding and transmission sequences are shown in**Table 1**. The encoding and also transmission sequences for the half-rate three transmit antennas are presented in **Table 2**. The transmission model is based on transmitting signals $C_t^i, i = 1,2,\ldots,n$ simultaneously from the n transmit antennas at each time slot of t, under the assumption that the wireless channel is quasi-static (so that the path gains are considered to be constant over a frame of length l and vary from one frame to the other one), and flat fading [3]. Defining the path gain from transmit antenna i to the receive antenna j by $\alpha_{i,j}$ and also denoting the noise samples by γ_t^j (i.e. independent samples of a zero mean complex Gaussian random variable with variance n/(2SNR) per complex system dimension), the received radio signal at the system antenna j at the time t is given as Equation (2) [3]:

$$\xrightarrow{yields} r_t^j = \sum_{i=1}^{n} \alpha_{i,j} c_t^i + \gamma_t^j . \tag{2}$$

The system's average energy of each transmitter antenna's symbols is normalized to one so that the average power of each receive antenna's radio signal is n. The receiver computes the decision metric given by the (3) over all the codewords, under the assumption that prefect channel impulse response (CSI) is available, and decides in favor of codeword that minimizes the sum [3]. The received signal is processed first in the channel estimator and combiner, and the combined received radio signals are then sent out to the system ML detector. In order for the wireless system to detect symbols of the two transmit antennas; denoted by s_1 and s_2; the decision metrics [36] have been potentially employed. For detecting symbols of half-rate three transmit antennas; denoted by s_1, s_2, s_3; the decoder minimizes the decision metrics, as in [36].

$$\xrightarrow{yields} \sum_{t=1}^{l}\sum_{j=1}^{m}\left| r_t^j - \sum_{i=1}^{n}\alpha_{i,j}c_t^i \right|^2 . \tag{3}$$

Table 1. Terrestrial Encoding and Transmission Sequence for Two Transmit Antennas; Alamouti Scheme; Code Rate: 1.

Time Slot	Antenna-I	Antenna-II
I	x_1	x_2
II	$-x_2^*$	x_1^*

Table 2. Terrestrial Encoding and Transmission Sequence for Three Transmit Antennas; Tarokh Scheme; Code Rate: 1/2.

Time Slot	Antenna-I	Antenna-II	Antenna-III
I	x_1	x_2	x_3
II	$-x_2$	x_1	$-x_4$
III	$-x_3$	x_4	x_1
IV	$-x_4$	$-x_3$	x_2
V	x_1^*	x_2^*	x_3^*
VI	$-x_2^*$	x_1^*	$-x_4^*$
VII	$-x_3^*$	x_4^*	x_1^*
VIII	$-x_4^*$	$-x_3^*$	x_2^*

6. TERRESTRIAL MIMO CHANNEL MODELING

From system level aspect MIMO channel is represented by a N × M matrix H; the received terrestrial radio signal $y=[y_1, y_2,..., y_m]^T$ can also be expressed as [37]:

$x=[x_1, x_2,..., x_m]^T$ $y=H\cdot x+\hat{n}=\sum_{m=1}^{M} h_m \cdot x_m+\hat{n}$ is the transmitted vector, $\hat{n}$ is the N × 1 additive white Gaussian noise (AWGN). If the number of receive antennas is reduced to one (i.e. for MISO systems), the MISO's CIS can then be considered as N SISO channels expressed as 1 × N vector. It is assumed that n_t transmit and n_r receive antennas are placed in uniform linear arrays of normalized lengths L_t and L_r respectively. The normalized separation between transmit antennas $\Delta_t$and the normalized separation between the receive antennas Δ_r are given: $\Delta_t=L_t/n_t$ and $\Delta_r=L_r/n_r$ respectively; where the system normalization is by the wavelength of the passband transmitted radio signal λ_c [37]. It is then supposed there is an arbitrary number of physical paths between the transmitter and receiver, and the attenuation of path $\aleph$, and its angles with the transmit and receive system array are denoted as: $a_\aleph$, $\varphi_{t\aleph}(\omega_{t\aleph}:=\cos\varphi_{t\aleph})$, and $\varphi_{r\aleph}(\omega_{r\aleph}:=\cos\varphi_{r\aleph})$. Channel matrix H is given as the following equations; where $d^{(i)}$ denotes the distance between the system transmit antenna 1 and the receive antenna 1, along the path $\aleph$ in radio transmission [37].

$$\underrightarrow{yields}\; H=\sum_{\aleph} a_\aleph^b e_r(\omega_{r\aleph}) e_t(\omega_{t\aleph})^*; \tag{4}$$

$$\underrightarrow{where}\; a_\aleph^b := a_\aleph \sqrt{n_t n_r}\exp\left(-\frac{j2\pi d^{(i)}}{\lambda_c}\right). \tag{5}$$

$$e_r(\omega):=\frac{1}{\sqrt{n_r}}\begin{bmatrix} 1 \\ \exp(-j2\pi\Delta_r\omega) \\ \vdots \\ \exp(-j2\pi(n_r-1)\Delta_r\omega) \end{bmatrix}. \tag{6}$$

$$e_t(\omega) := \frac{1}{\sqrt{n_t}} \begin{bmatrix} 1 \\ \exp(-j2\pi\Delta_t\omega) \\ \vdots \\ \exp(-j2\pi(n_t - 1)\Delta_t\omega) \end{bmatrix}. \quad (7)$$

The column vectors: $e_t(\omega)$ and $e_r(\omega)$ are the transmitted and received unit spatial signatures along direction ω respectively. There are two primary applications for the channel models: location-specific models which provide very accurate models robust with respect to small errors, for optimizing a system in a certain geographical region (for the planning and deployment). Simplified channel models that describe the CIR statistics in a parametric form; in which the number of parameters is small, and is independent of the specific locations. The models which are reflecting the important properties of propagation channels are used for the design, testing, and type approval of wireless systems [38]. For comparing different MIMO systems and their relevant PHY layer performance enhancement techniques in a manner that is unified and agreed on by many system parties, various organizations introduce the reference channel models with reproducible conditions [38]. Among these modeling methods utilized in site-specific and generic channel model applications (i.e. stored CIR, deterministic channel models, and also stochastic system channel models), the latter one is then selected which describes the predicted probability density function (PDF) of the CIR over a large area. In this research, stochastic modeling of the received radio signal's amplitude is described by Rayleigh distribution which is used model for predicting PDF of the CIR; in the urban areas with blocked line-of-sight (LoS). Meanwhile, the COST207 models proposed bivariate Gaussian Doppler spectrum for modeling long echoes in urban propagation environments (models in the urban areas) [38]; given by:

$$P_G(f) = \left(1 / \sum_{i=1}^{2} A_i\right)\left[\sum_{i=1}^{2}\left(A_i / \sqrt{2\pi\sigma_i^2}\right) \cdot \exp\left(-(f - f_i)^2 / 2\sigma_i^2\right)\right];$$

where f_i, A_i, and σ_i, are center frequencies, power gains, and standard deviations respectively [35,39,40].

7. TERRESTRIAL TDE MODELING AND ANALYSIS

In the case of complex and real amplitude error functions and in practice, the samples of the received and transmitted signals are existed for the time interval 0 < t < T. The received terrestrial waveform can be modeled as (8):

$$r(nT_s)=\sum_{k=1}^{M}a_k s(nT_s-\tau_k)+\omega[n];0\le n\le N-1. \tag{8}$$

where the sampling interval $T_s = T/N$ and $\omega[n]$s are the noise samples of the AWGN with the terrestrial wireless system continuous autocorrelation as (9):

$$\xrightarrow{\text{yields}} R_{\tilde{\omega}}(\tau)=\sigma^2{}_{\tilde{\omega}}\delta(\tau). \tag{9}$$

where it is assumed that $\tilde{\omega}(t)$ is low-pass filtered (LPF) by the factor as in the Equation (10) below, in order to yield $\tilde{\omega}(t)$ such that the discrete system radio signal has a flat power spectrum as (11).

$$h_{L-P}(t)=\frac{1}{\pi t}\sin\left(\frac{\pi t}{T_s}\right)\stackrel{\tilde{\omega}(t)}{\Rightarrow}\omega[n]=\tilde{\omega}(nT_s); \tag{10}$$

$$\xrightarrow{\text{yields}} \sigma^2{}_{\tilde{\omega}}\equiv E\{\omega^2[n]\}=\frac{\sigma^2{}_{\tilde{\omega}}}{T_s}. \tag{11}$$

The least squares estimator (LSE) of $a = [a_1, a_2,\cdots, a_M]^T$ and $\tau = [\tau_1, \tau_2,\cdots, \tau_M]^T$ has then been obtained by minimizing the squared error function (12) as part of the system time-delay estimation (TDE) computation.

$$E_r(\tau,a)=\sum_{n\in N}\left|R[n]-S[n]\sum_{k=1}^{M}a_k e^{\frac{-j2n\pi\tau_k}{NT_s}}\right|^2;$$
$$\xrightarrow{\text{yields}} \left\|\tilde{r}-\tilde{p}(\lambda)a^2\right\|. \tag{12}$$

$$N=\left\{0\le n\le\frac{N}{2}-1;S[n]>threshold\right\}=\{q_1,\cdots,q_L\}. \tag{13}$$

where R[n] and S[n] are the discrete Fourier transforms (DFTs) of r[nT$_S$] and s[nT$_S$], respectively. In the presence of AWGN, the LSE is also the maximum likelihood estimator (MLE). In the minimization procedure, a threshold is specified in order to employ only the spectrum regions with the satisfactory system SNRs. The weighting function is employed in order to enhance the accuracy of the estimated delay by attenuating the radiowaves fed into the correlator in spectral region where the SNR is the lowest. For the uncorrelated Gaussian radio system signal, noise, and single path, TDE is asymptotically unbiased and efficient in terms of the limit of long observation intervals. The terrestrial weights the cross-spectral phase according to the estimated phase when variance of the estimated phase error is the least. The system TDE procedure adapts the filter in order to insert a delay equal and opposite to that existing between the two radio signals. In the case of the AWGN presence, the filter weight corresponding to true-delay obtains its maximum value compared to the other TDE terrestrial wireless system filter weights.

$$\lambda = [\lambda_1, \lambda_2, \cdots, \lambda_M]^T; \lambda_k = \frac{-2\pi\tau_k}{NT_s} \xrightarrow{yields}$$
$$r = [R[q_1], R[q_2], \cdots, R[q_L]]^T;$$
$$S = diag(S[q_1], \cdots, S[q_L]). \tag{14}$$

$$A(\lambda) = \begin{bmatrix} e^{j\lambda_1 q_1} & e^{j\lambda_2 q_1} & \cdots & e^{j\lambda_{M-1} q_1} & e^{j\lambda_M q_1} \\ e^{j\lambda_1 q_2} & e^{j\lambda_2 q_2} & & e^{j\lambda_{M-1} q_2} & e^{j\lambda_M q_2} \\ \vdots & & \ddots & & \vdots \\ e^{j\lambda_1 q_L} & e^{j\lambda_2 q_L} & \cdots & e^{j\lambda_{M-1} q_L} & e^{j\lambda_M q_L} \end{bmatrix};$$
$$\xrightarrow{yields} \tilde{p}(\lambda) = SA(\lambda); \tilde{p}(\lambda) = [p(\lambda_1), \cdots, p(\lambda_M)]. \tag{15}$$

where in the relation (14), S[q$_k$] and R[q$_k$] are the timedelay samples that lie above the specified terrestrial wireless system time-delay estimation threshold.

$$\tilde{P}(\lambda) = Q(\lambda)R(\lambda) \overset{QR\ Decomposition}{\Rightarrow}$$
$$Q(\lambda) = [q_1(\lambda_1), q_2(\lambda_1, \lambda_2), \cdots, q_M(\lambda_1, \cdots, \lambda_M)]. \tag{16}$$

where in the relation (16), the q_is are orthogonal, and each one is a function of $(\lambda_1, \lambda_2, \cdots, \lambda_i)$. Hence $E_r(\lambda)$ can be written as Equation (17) below, in order to minimize the error function for the terrestrial radio signal time-delays.

$$E_r(\lambda) = \left\| \left[I - \tilde{p}\left(\tilde{p}^H \tilde{p}\right)^{-1} \tilde{p}^H \right] \tilde{r} \right\|^2;$$

$$\overset{QR\ Decomposition}{\Rightarrow} E_r(\lambda) = \left\| \left[I - QQ^H \right] \tilde{r} \right\|^2.$$

(17)

8. TERRESTRIAL COMMUNICATION RESULTS

In this section the simulations are presented which depict the performance (BER) of the terrestrial wireless system. For these simulations, a rate 1/3 SCC has been employed, and formed by the unpunctured recursive convolutional component codes with the polynomial matrices for outer and inner codes. The re-ordering of the outer code's output is performed based on the algorithm presented in [30] which shown to provide additional coding gains for the SCCs. In this work, the interleaver size is 4096. For the simulations, the wireless communication system employs 16-ary QAM modulation with gray coded bit mapping, and it is assumed that the additive noise at the receiver is AWGN. **Figure 2** presents the performance for the discussed wireless system of interest with two transmitter and one receiver antennas (i.e. diversity order $D_O = 2 \times 1 = 2$); and the system results have been compared with the BER curves associated with the theoretical system performance of the uncoded 16-QAM modulation in the Rayleigh multipath channel with $D_O = 1$ (i.e. SISO without diversity), and $D_O = 2$. The figure presents a good terrestrial system reference for comparison since in both approaches, the channel is assumed to be highly faded with blocked LoS. Hence, the additional coding of more than 15 dB implies that the employed technique yields astonishing additional coding gains at the expense of the tolerable system decoding complexities. Meanwhile, it should also be noted that the slope of the BER curve could be simply increased by increasing the component code's constraint length, the number of decoding iterations, and the interleaver size, at the expense of the delays which might not be acceptable for the hard real-time systems. In addition one more antenna at the transmitter and receiver has been employed in order to evaluate the coding gains due to the deployment of an additional antenna, while relaxing the other system simulation parameters for the designed SCC. In this system Tarokh scheme STBCs have been utilized ($D_O = 3 \times 2 = 6$), and compared with theoretical uncoded 16-QAM system over the Rayleigh faded channel with $D_O = 6$. As it can be seen from the figure, while the former scheme results in the BER of 10^{-5} at SNR = 9 dB, the latter yields the BER of 10^{-7}, which should be considered as an immense terrestrial system advancement. Meanwhile, the introduced enhancement strategy to the wireless system is confirmed from the scatter diagrams (i.e. **Figure 3**) of the received constellation shapes which have shown that although the received symbols are highly faded for low the SNRs, they land at almost the same locations as in the transmitted constellations (these symbols are transmitted on ±1 + ±j, ±1 + ±3j, ±3 + ±3j and ±3j + ±1j). Figures 4 and 5 present cross-correlation with sharp peak at the correct terrestrial time-delay with the time-lag corresponding to the minimum error as the actual time-delay, and the proposed relation between the SNR and the estimated timedelay in terrestrial environment, respectively.

9. CONCLUDING REMARKS

In this contribution, the performances of space-time codes based on Alamouti and Tarokh STBCs have been thoroughly evaluated in conjunction with high-performance turbo coding system schemes which were previously developed in order to yield astonishing system coding gains.

The introduced technique in this research could also be considered as a potential promising candidate for futuristic terrestrial systems and applications; wherein the channel environment is highly faded, and power and bandwidth are at premium so very low and high-modulation orders may not be adopted for the common highly-faded conditions of the terrestrial propagation environments.

As for the future research, efforts will be put into considering the proposed technique combined with the TCM for improving the system's efficiency. In this contribution, a promising mathematically-oriented analytical evaluation for the terrestrial system time-delay estimation (TDE) in presence of AWGN in the multipath environment has also been carried out. The presented analytical model can further be extended in order to analyze the terrestrial wireless communication system overall performance for the unknown number of radio paths in specified environment.

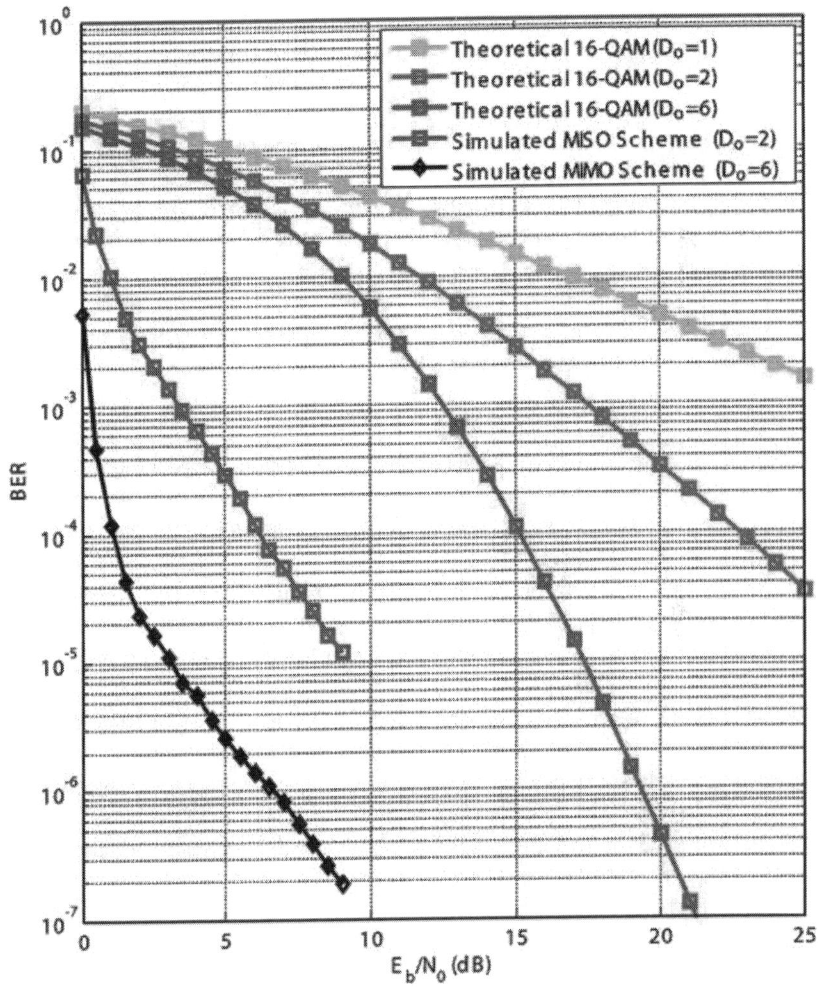

Figure 2. Comparison of BER vs. (E_b/N_0) for coded 16-ary QAM MISO (2-Tx, 1-Rx) and MIMO (3-Tx, 2-Rx) terrestrial communication systems.

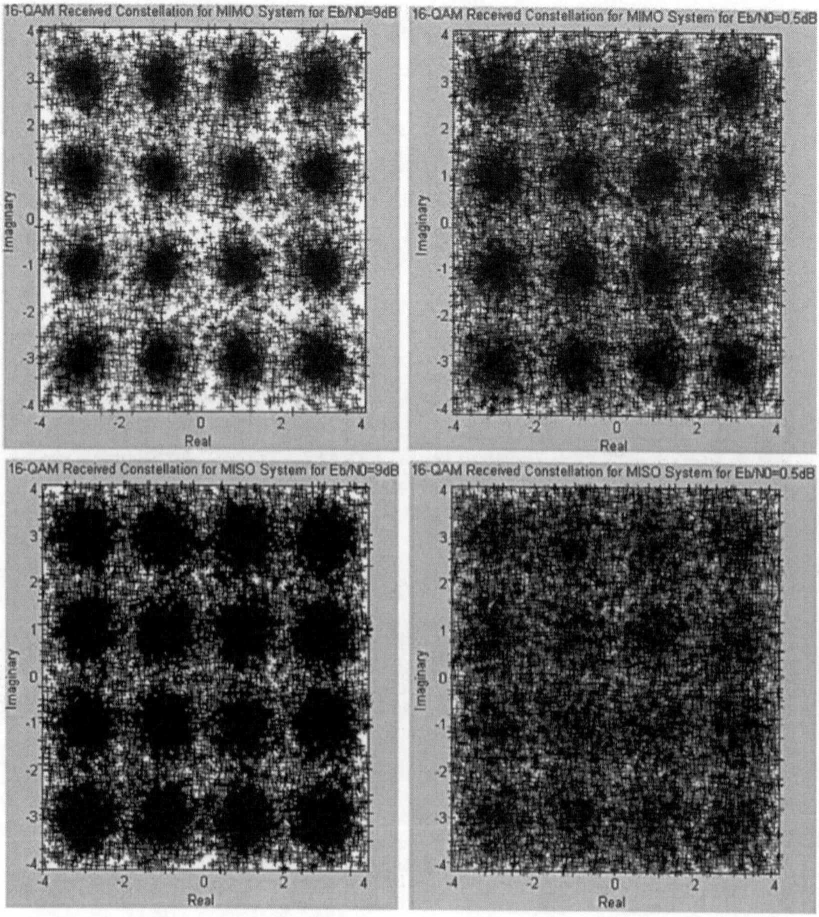

Figure 3. Comparison of constellations at the MISO and MIMO wireless terrestrial communications receivers for M = 16-ary QAM digital system modulation scheme with gray coded bit assignment.

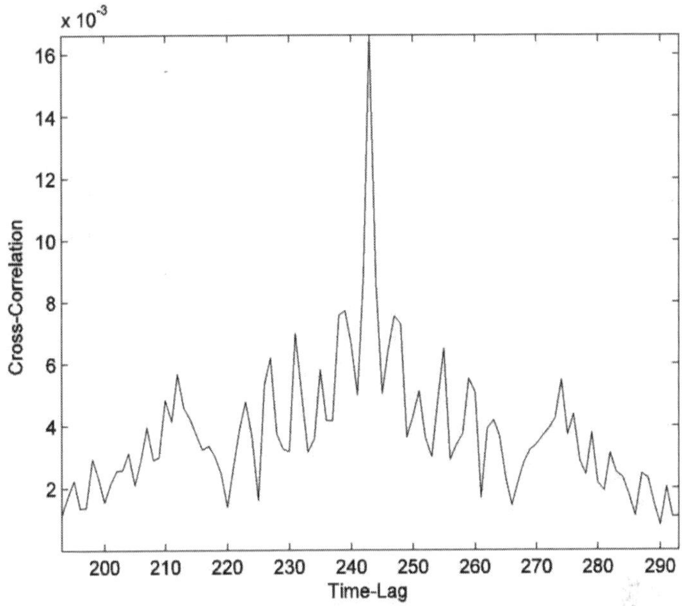

Figure 4. Cross-correlation based on MLE in a simulated multipath terrestrial communications environment in presence of AWGN.

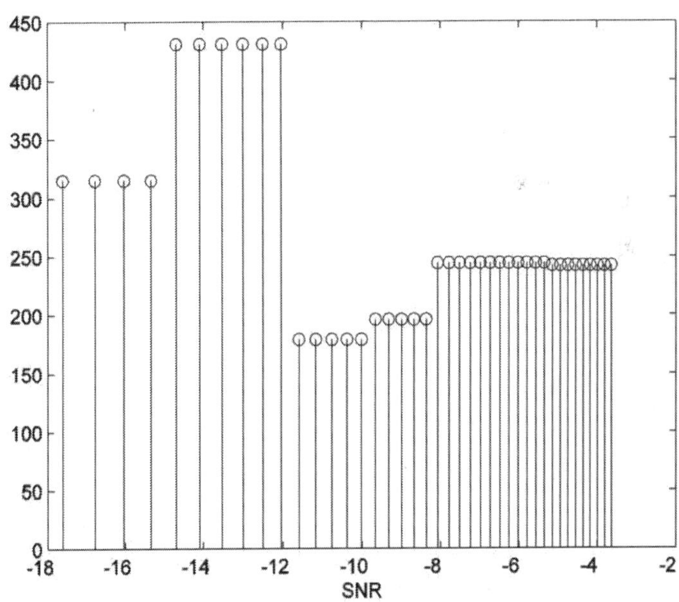

Figure 5. The relation between SNR and estimated timedelays using MLE in a simulated multipath terrestrial communications system environment in presence of AWGN.

REFERENCES

1. S. M. Alamouti, "A Simple Transmit Diversity Technique for Wireless Communications," IEEE Journal on Selected Areas in Communications, Vol. 16, No. 8, 1998, pp. 1451-1458.

2. V. Tarokh, H. Jafarkhani and A. R. Calderbank, "SpaceTime Block Codes from Orthogonal Designs," IEEE Transactions on Information Theory, Vol. 45, 1999, pp. 1456-1467.

3. V. Tarokh, H. Jafarkhani and A. R. Calderbank, "SpaceTime Block Coding for Wireless Communications: Performance Results," IEEE Journal on Selected Areas in Communications, Vol. 17, No. 3, 1999, pp. 451-460.

4. T. H. Liew and L. Hanzo, "Space-Time Codes and Concatenated Channel Codes for Wireless Communications," Proceedings of the IEEE, Vol. 90, No. 2, 2002, pp. 187- 219.

5. X. Zhou, X. Quan and R. Li, "A Dual-Broadband MIMO Antenna System for GSM/UMTS/LTE and WLAN Handset," IEEE Antennas and Wireless Propagation Letters, Vol. 11, 2012, pp. 551-554.

6. W. Jaafar, W. Ajib and S. Tabbane, "The Capacity of MIMO-Based Wireless Mesh Networks," Proceedings of International Conference on Neural Networks (ICON 2007), Adelaide, 19-21 November 2007, pp. 259-264.

7. Q. Li, X. Lin, and J. Zhang, and W. Roh, "Advancement of MIMO Technology in WiMAX: From IEEE 802. 16d/e/j to 802.16m," IEEE Communications Magazine, Vol. 47, No. 6, 2009, pp. 100-107.

8. R. Langwieser, C. Angerer and A. L. Scholts, "A UHF Frontend for MIMO Applications in RFID," 2010 IEEE Radio and Wireless Symposium (RWS), New Orleans, 10-14 January 2010, pp. 124-127.

9. C. Berrou, A. Glavieux and P. Thitimajshima, "Near Shannon Limit Error-Correcting Coding: Turbo Codes," Proceedings of IEEE International Conference on Communications, 1993, pp. 1064-1070.

10. D. Zhao, L. Yue and J. Yang, "Application of Turbo Codes in High-Speed Real-Time Channel," Journal of Electronics (China), Vol. 23, No. 4, 2006, pp. 602-605.

11. ETSI, "Digital Video Broadcasting (DVB). Interaction Channel for Satellite Distribution Systems," DVB-RCS 001, 2000.

12. 3rd Generation Partnership Project (3GPP), "Multiplexing and Channel Coding (FDD)," 3G TS 25.212, 1999.

13. J. Zyren, "Overview of the 3GPP Long Term Evolution Physical Layer," Freescale Semiconductor Technical Report, 2007.

14. A. Ebrahimi and A. Rahimian, "Estimation of Channel Parameters in a Multipath Environment via Optimizing Highly Oscillatory Error Functions Using a Genetic Algorithm," 15th International Conference on Software, Telecommunications and Computer Networks (SoftCOM), Split-Dubrovnik, 15-17 September 2007, pp. 1-5.

15. P. P. Moghaddam, H. Amindavar and R. L. Kirlin, "A New Time-Delay Estimation in Multipath," IEEE Transactions on Signal Processing, Vol. 51, No. 5, 2003, pp. 1129-1142.

16. H. Amindavar and A. M. Reza, "A New Simultaneous Estimation of Directions of Arrival and Channel Parameters in a Multipath Environment," IEEE Transactions on Signal Processing, Vol. 53, No. 2, 2005, pp. 471-483.

17. H. C. So, "Analysis of an Adaptive Algorithm for Unbiased Multipath Time Delay Estimation," IEEE Transactions on Aerospace and Electronic Systems, Vol. 39, No. 3, 2003, pp. 777-785.

18. M. M. Kamruzzaman and Li Hao, "Performance of Turbo-SISO, Turbo-SIMO, Turbo-MISO and TurboMIMO System Using STBC," Journal of Communications—Academy Publisher, Vol. 6, No. 8, 2011, pp. 633- 639.

19. A. Hjorungnes and D. Gesbert, "Precoded Orthogonal Space-Time Block Codes over Correlated Ricean MIMO Channels," IEEE Transactions on Signal Processing, Vol. 55, No. 2, 2007, pp. 779-783.

20. F. Jiang, E. Psota and L. C. Perez, "Decoding Turbo Codes Based on Their Parity-Check Matrices," Proceedings of 39th Southeastern Symposium on System Theory, 2007, pp. 221-224.

21. J. Y. Kim and H. V. Poor, "Turbo-Coded Optical DirectDetection CDMA System with PPM Modulation," Journal of Lightwave Technology, Vol. 19, No. 3, 2001, pp. 312-323.

22. L. Bahl, J. Cocke, F. Jelinek and J. Raviv, "Optimal Decoding of Linear Codes for Minimizing Symbol Error Rate," IEEE Transactions on Information Theory, Vol. 20, No. 2, 1974, pp. 284-287.

23. W. Koch and A. Baier, "Optimum and Sub-Optimum Detection of Coded Data Distributed by Time-Varying Intersymbol Interference," Proceeings of IEEE GLOBECOM, Vol. 3, 1990, pp. 1679-1684.

24. P. Robertson, E. Villebrun and P. Hoeher, "A Comparison of Optimal and Sub-Optimal MAP Decoding Algorithms Operating in the Log Domain," Proceedings of IEEE International Conference on Communications (ICC), Vol. 2, 1995, pp. 1009-1013.

25. J. Erfanian, S. Pasupathy and P. G. Gulak, "Reduced Complexity Symbol Detectors with Parallel Structure for ISI Channels," IEEE Transactions on Communications, Vol. 42, No. 234, 1994, pp. 1661-1671.

26. S. Benedetto and G. Montorsi, "Serial Concatenation of Block and Convolutional Codes," Electronics Letters, Vol. 32, No. 10, 1996, pp. 887-888.

27. S. Benedetto, D. Divsalar, G. Montorsi and F. Pollara, "Serial Concatenation of Interleaved Codes: Performance Analysis, Design, and Iterative Decoding," The Telecommunications and Data Acquisition Progress Report, Jet Propulsion Lab, 1996, pp. 1-26,

28. F. Mehran, "Interleaver Choice in Serially Concatenated Codes for Soft Real-Time Wireless Speech Transmission Applications," Proceedings of 19th Iranian Conference on Electrical Engineering (ICEE), Tehran, 17-19 May 2011, pp. 1-6.

29. O. Y. Takeshita and D. Costello, "New Classes of Algebraic Interleavers for Turbo-Codes," IEEE International Symposium on Information Theory (ISIT), Cambridge, 16-21 August 1998.

30. F. Mehran and A. Rahimian "Physical Layer Performance Enhancement for Femtocell SISO/MISO Soft Real-Time Wireless Communication Systems Employing Serial Concatenation of Quadratic Interleaved Codes," Proceedings of 20th Iranian

Conference on Electrical Engineering (ICEE), May 2012, pp. 1188-1193.

31. O. Y. Takeshita and D. J. Costello, "New Deterministic Interleaver Designs for Turbo Codes," IEEE Transactions on Information Theory, Vol. 46, No. 6, 2000, pp. 1988- 2006.

32. H.-Q. Lai and K. J. R. Liu, "Space-Time Network Coding," IEEE Transactions on Signal Processing, Vol. 59, No. 4, 2011, pp. 1706-1718.

33. E. Telatar, "Capacity of Multi-Antenna Gaussian Channels," European Transactions on Telecommunications, Vol. 10, No. 6, 1999, pp. 585-595.

34. G. J. Foschini and M. J. Gans, "On Limits of Wireless Communications in a Fading Environment When Using Multiple Antennas," Wireless Personal Communications., Vol. 6, No. 2, 1998, pp. 311-335.

35. A. Rahimian and F. Mehran, "RF Link Budget Analysis in Urban Propagation Microcell Environment for Mobile Radio Communication Systems Link Planning," Proceedings of International Conference on Wireless Communications and Signal Processing (WCSP), Nanjing, 9-11 November 2011, pp. 1-5.

36. A. Rahimian and F. Mehran, "BEP Enhancement for Semi-Femtocell MIMO Systems Employing SC-QICs and OSTBCs," International Journal of Electronics Communication and Computer Technology (IJECCT), Vol. 3, No. 1, 2013, pp. 329-332.

37. D. Tse and P. Viswanath, Fundamentals of Wireless Communication, Cambridge University Press, Cambridge, 2005.

38. A. F. Molisch, Wireless Communications, 2nd Edition, John Wiley & Sons Ltd., Wiley-IEEE Press, New York, 2011.

39. P. Almers, et al., "Survey of Channel and Radio Propagation Models for Wireless MIMO Systems," EURASIP Journal on Wireless Communications and Networking, 2007.

40. J.-M. Conrat and P. Pajusco, "Directional Propagation Channel Estimation and Analysis in Urban Environment with Panoramic Photography," International Journal of Microwave and Wireless Technologies, Vol. 4, No. 1, 2012, pp. 3-13.

8

ROBUST SLIDING MODE CONTROL FOR NONLINEAR DISCRETE-TIME DELAYED SYSTEMS BASED ON NEURAL NETWORK

Vishal Goyal, Vinay Kumar Deolia, Tripti Nath Sharma

Department of Electronics and Communication Engineering,
G. L. A. University, Mathura, India

ABSTRACT

This paper presents a robust sliding mode controller for a class of unknown nonlinear discrete- time systems in the presence of fixed time delay. A neural-network approximation and the Lyapunov-Krasovskii functional theory into the sliding-mode technique is used and a neural-network based sliding mode control scheme is proposed. Because of the novality of Chebyshev Neural Networks (CNNs), that it requires much less computation time as compare to multi layer neural network (MLNN), is preferred to approximate the unknown system functions. By means of linear matrix inequalities, a sufficient condition is derived to ensure the asymptotic stability such that the sliding mode dynamics is restricted to the defined sliding surface. The proposed sliding mode control technique guarantees the system state trajectory to the designed sliding surface. Finally, simulation results illustrate the main characteristics and performance of the proposed approach.

KEYWORDS

Discrete-Time Nonlinear Systems, Lyapunov-Krasovskii Functional, Linear Matrix Inequality (LMI), Sliding Mode Control (SMC), Chebyshev Neural Networks (CNNs)

1. INTRODUCTION

Time delay is undesirable parameter which is often encountered in various engineering systems, such as mechanical systems, chemical systems, and so on. The time delay degrades the system performance and leads to instability of the system. As a design tool for robust motion control system, SMC has been well designed for a wide range of nonlinear systems in both continuous time and discrete time. SMC is robust to parametric uncertainties and insensitive to unknown disturbance. SMC has been studied in seventies by the name of variable structure control (VSC) [1] [2] . VSC, is characterized by discontinuous feedback control law which switches the system in a predefined subspace [3] . Its implementation by a digital system requires sampling interval which leads to chattering. Over the past few decades, considerable attention has been reported to the stability analysis of continuous time delay systems by using different approaches [4] - [10] . In [4] , robust controller has been designed for continuous time delay system using Ricacati equation approach. Considerable, attention has been given using LMI approach for stabilization of continuous time delay systems [5] - [8] . An adaptive control approach has been proposed for the control of time delay system [9] [10] . A discrete SMC (DSMC) is important when we implement robust control digitally with slow sampling rate. It is important to note that DSMC cannot be obtained from its continuous counterpart by simple conversion. Since modeling inaccuracy and external conditions lead to uncertainties, disturbances and nonlinearities in systems. Hence, the stability analysis of uncertain discrete-time delay systems have been studied over past few years with different control approaches have been well documented in [11] and reference therein. Moreover, in the above papers, the unknown nonlinearities have not been investigated. In [12] , a robust control of uncertain nonlinear state delayed system, which gives a conservative condition of control, is presented. In recent years, many papers have reported the problem of SMC for state delay uncertain systems [13] - [17] . Most of these papers for uncertain time delay systems involve norm- bound nonlinearities which are treated as external disturbances. Adaptive multilayer neural control schemes for the control of complex nonlinear systems have shown great results over past few years. Now, it is an established fact that unknown nonlinear functions can be approximated from neural network. Neural network appears a powerful tool for nonlinear control problems [18] - [20] . In [21] , the SMC have been used for control of uncertain state-delay system with unknown nonlinearity. In this work, Chebyshev Neural Network is used to estimate the unknown nonlinearity and linear matrix inequalities (LMI) conditions were derived to ensure the asymptotic stability on the defined sliding surface. Adaptive SMC for a class of discrete nonlinear systems was proposed. The proposed controller uses switching function with adaptive term to reduce the problem of chattering. Artificial neural network (ANN) was used for approximation of modeling errors. The nonlinearity is strictly positive and bounded away from zero [22] . In [23] , a new SMC has been used to control the unknown nonlinear discrete-time systems. The chattering is reduced as compared to normal discrete-time sliding mode control using time varying gain.

This paper proposes a discrete-time sliding mode controller for a class of state delay nonlinear discrete systems. The unknown nonlinear functions in system dynamics is approximated using Chebyshev Neural Networks (CNNs). New weight update laws are derived to make this scheme adaptive. The stability of state delay system is taken care by carefully selecting Lyapunov-Krasovskii functional candidate. Thus conservative, sufficient conditions were derived which was represented by an appropriate set of LMIs.

The paper organization is as follows. Section 2 presents the CNN structure. Problem formulation and preliminaries are elaborated in Section 3. Section 4 presents controller design is stated in detail. The stability analysis is presented in Section 5. The effectiveness of proposed scheme is validates through simulation results in Section 6. The note ends with concluding remarks in Section 7.

Notations: $\|\cdot\|$ denotes Euclidean norm, $\|\cdot\|_F$ implies Frobenius norm. The $\mathrm{tr}(\cdot)$ stands for trace of matrix.

2. CNN STRUCTURE

An ANN is a simple interconnected group of nonlinear elements, which has the capability to represents nonlinear functions. The representation accuracy depends on the ANN complexity, i.e., the number of elements and the way in which they are interconnected [24] .

There is different ANN configuration available, like feed forward network such as multilayer perceptron (MLP), radial basis function (RBF) networks, Chebyshev neural network (CNN) etc. The MLP network has certain disadvantage that it requires a large amount of computation for learning. The RBF network can effectively learn from discontinuities and local variations. The problem with this network is choosing an appropriate set of RBF centers for effective learning. A single-layer functional link artificial neural network (FLANN) in which the need of hidden layer is eliminated by expanding the input pattern using Chebyshev polynomials. The main advantage of this network is that it requires much less computation as compared to a multilayer perceptron (MLP).

CNN is a functional link network (FLN) based on Chebyshev polynomials. CNN architecture has two main parts, namely, numerical transformation and learning [25] . In numerical transformation we use finite set of Chebyshev polynomials as a functional expansion (FE) of input pattern. The learning part is a functional-link neural network based on Chebyshev polynomials. The Chebyshev polynomials can be obtained by a recursive formula

$$T_{i+1}(x) = 2xT_i(x) - T_{i-1}(x), \quad T_0(x) = 1 \tag{1}$$

where, $T_1(x)$ are Chebyshev polynomials, i is the order of polynomials chosen and here x is a scalar quantity. The different choices of $T_1(x)$ are $x \& 2x$.

The output of single layer neural network is given by

$$\hat{g}(x) = \hat{w}^{\mathrm{T}}\phi \tag{2}$$

where, w are the weights and ϕ is the suitable basis function of neural network. Based on the approximation property of CNN [27] -[30] , there exist ideal weights W, so that the function $g(x)$ to be approximated can be represented as

$$g(x) = w^{\mathrm{T}}\phi + \varepsilon \tag{3}$$

where, ε is the CNN functional reconstruction error vector and $\|\varepsilon\| \leq \varepsilon_N$ is bounded.

3. PROBLEM FORMULATION

Consider the following discrete-time state delay system as in [26] (Figure 1)

$$x(k+1) = Ax(k) + A_d x(k-h) + g(x(k))u(k) \tag{4}$$

where, $x(k) \in R^n$ and $u(k) \in R^m$ denote the state and input vectors respectively. A and A_d are real constant matrices with appropriate dimensions. $g(x(k))$ is a unknown nonlinear function of a given system in Equation (4), and h is a positive number representing delay.

For the system given in (4) the sliding mode controller is obtained as

$$u(k) = -\frac{1}{\hat{g}(x(k))}\left[Ax(k) + A_d x(k-h)\right] \tag{5}$$

where $\hat{g}(x(k))$ is the approximated value of the nonlinear function.

The objective of this work is to guarantee the stability of sliding mode controller in Equation (5) of the nonlinear system Equation (4), so that the system stays on the sliding surface.

4. CONTROLLER DESIGN

The first step in the design of discrete-time SMC control algorithm would be the design of sliding surface. The linear sliding surface is defined as:

$$s(k) = Cx(k) \tag{6}$$

where $C \in R^{m \times n}$ is a real matrix of appropriate dimensions.

For a system to be asymptotically stable, the sliding surface is defined as follows.

$$s(k) = 0 \tag{7}$$

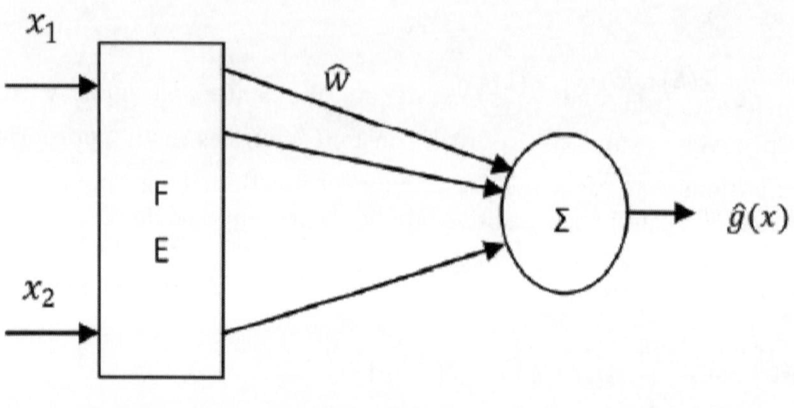

Figure 1. Chebyshev neural network.

The second step is to design a control law which can guarantee the sliding mode reaching condition of the given linear sliding surface. The obtained control law is given in Equation (5) will force the trajectory of the system to move towards the sliding surface monotonically and causes zigzag motion around the sliding surface.

5. STABILITY ANALYSIS

The following assumptions are needed for the stability analysis of the given unknown nonlinear system [26] .

Assumption 1: The state delay h is a constant time delay that is basically induced by the network transmission. For constant time delay the lower and upper bounds are assumed to be identical.

Assumption 2: The nonlinear function $g(x(k))$ in the system is unknown and bounded.

Assumption 3: (Bounded Ideal NN Weights): The ideal NN weights w are bounded so that $\|w\| \le w_M$, with w_M a known bound. The symbol $\|\cdot\|_F$ denotes the Frobenius norm, i.e. given a matrix A, the Frobenius norm is given by,

$$\|A\|_F^2 = \mathrm{tr}\left(A^{\mathrm{T}}A\right)$$

Assumption 4: Let $\tilde{g}(x(k)) = G\hat{g}(x(k))$, where $G = G^{\mathrm{T}}$ is a $n \times n$ symmetric matrix, and $\tilde{g}(x(k))$ and $\hat{g}x(k)$ are the n-column vectors.

Theorem 1:

Given the system in Equation (4) and Assumptions 1 - 4, sliding mode control law Equation (5), the estimated NN weights are given by

$$\hat{w}(k+1) = \left[\left(\hat{w}(k)\right) + \left(x^{\mathrm{T}}(k)(Q-P-Z)x(k)\right)^{1/2} + \left(\hat{M}^{\mathrm{T}}\left(x(k)\right)Z\hat{M}\left(x(k)\right)\right)^{1/2}\right] \tag{8}$$

with the condition are

$$4\left(\left(\hat{w}(k)\right)\left(\hat{M}^{\mathrm{T}}\left(x(k)\right)Z\hat{M}\left(x(k)\right)\right)^{1/2}\left(x^{\mathrm{T}}(k)(Q-P-Z)x(k)\right)^{1/2}\right)^{1/2} > 0 \tag{9}$$

$$2\left(x^{\mathrm{T}}(k)(Q-P-Z)x(k)\right)^{1/2}\left(\hat{M}^{\mathrm{T}}\left(x(k)\right)Z\hat{M}\left(x(k)\right)\right)^{1/2} > 0 \tag{10}$$

Suppose there exist an $n\times n$ positive-definite matrix P, an $n\times n$ nonnegative-definite matrix Q, an $n\times n$ nonnegative-definite matrix z and $n\times n$ symmetric matrix G such that following LMI holds,

$$H1)=\begin{bmatrix} A^{\mathrm{T}}G^{\mathrm{T}}PGA-A^{\mathrm{T}}G^{\mathrm{T}}zGA-A^{\mathrm{T}}G^{\mathrm{T}}z-zGA & A^{\mathrm{T}}G^{\mathrm{T}}PGA_d-A^{\mathrm{T}}G^{\mathrm{T}}zGA_d-GzA_d \\ * & A_d^{\mathrm{T}}G^{\mathrm{T}}PGA_d-A_d^{\mathrm{T}}G^{\mathrm{T}}zGA_d-Q \end{bmatrix}<0 \tag{11}$$

Thus by properly selecting the control gain and the design parameters, the state trajectory is reaching on the designed sliding surface.

Proof: Choose Lyapunov-Krasovskii functional candidate,

$$V(k)=V_1(k)+V_2(k)+V_3(k)+V_4(k) \tag{12}$$

where

$$V_1(k)=x^{\mathrm{T}}(k)Px(k) \tag{13}$$

$$V_2(k)=\sum_{i=k-h}^{k-1}x^{\mathrm{T}}(i)Qx(i) \tag{14}$$

$$V_3(k)=\mathrm{tr}\left(\tilde{w}^{\mathrm{T}}(k)\tilde{w}(k)\right) \tag{15}$$

$$V_4(k)=\eta^{\mathrm{T}}(k)Z\eta(k) \tag{16}$$

And

$$\eta(k)=x(k+1)-x(k) \tag{17}$$

Substituting Equations (13)-(16) in Equation (12)

$$V(k)=x^{\mathrm{T}}(k)Px(k)+\sum_{i=k-h}^{k-1}x^{\mathrm{T}}(i)Qx(i)+\mathrm{tr}\left(\tilde{w}^{\mathrm{T}}(k)\tilde{w}(k)\right)+\eta^{\mathrm{T}}(k)Z\eta(k) \tag{18}$$

and

$$V(k+1)=x^{\mathrm{T}}(k+1)Px(k+1)+\sum_{i=k-h+1}^{k}x^{\mathrm{T}}(i)Qx(i)+\mathrm{tr}\left(\tilde{w}^{\mathrm{T}}(k+1)\tilde{w}(k+1)\right)+\hat{M}^{\mathrm{T}}(x(k))Z\hat{M}(x(k)) \tag{19}$$

where

$$\eta(k+1)=\hat{M}(x(k)) \tag{20}$$

Since P, Z is a positive-definite and Q is a nonnegative-definite, $V(k)$ is then positive-definite. Therefore,

$$\Delta V(k)=V(k+1)-V(k) \tag{21}$$

Substituting Equation (18) and (19) in Equation (21),

$$\begin{aligned}\Delta V(k)=&\,x^{\mathrm{T}}(k+1)Px(k+1)+\sum_{i=k-h+1}^{k}x^{\mathrm{T}}(i)Qx(i)+\mathrm{tr}\left(\tilde{w}^{\mathrm{T}}(k+1)\tilde{w}(k+1)\right)+\hat{M}^{\mathrm{T}}(x(k))Z\hat{M}(x(k))\\&-x^{\mathrm{T}}(k)Px(k)-\sum_{i=k-h}^{k-1}x^{\mathrm{T}}(i)Qx(i)-\mathrm{tr}\left(\tilde{w}^{\mathrm{T}}(k)\tilde{w}(k)\right)-\eta^{\mathrm{T}}(k)Z\eta(k).\end{aligned} \tag{22}$$

Substituting Equation (5) in Equation (21) and using Assumption 3 in Equation (22)

$$\begin{aligned}\Delta V(k)=&\left[Ax(k)+A_dx(k-h)+\tilde{g}(x(k))u(k)+\hat{g}(x(k))u(k)\right]\\&\times P\left[Ax(k)+A_dx(k-h)+\tilde{g}(x(k))\,u(k)+\hat{g}(x(k))u(k)\right]\\&+x^{\mathrm{T}}(k)Qx(k)+\hat{M}^{\mathrm{T}}(x(k))Z\hat{M}(x(k))-x^{\mathrm{T}}(k)Px(k)-x^{\mathrm{T}}(k-h)Qx(k-h)\\&-\left[x(k+1)-x(k)\right]^{\mathrm{T}}Z\left[x(k+1)-x(k)\right]+\left\|\tilde{w}^{\mathrm{T}}(k+1)\right\|^2-\left\|\tilde{w}^{\mathrm{T}}(k)\right\|^2.\end{aligned} \tag{23}$$

After some mathematical manipulations in Equation (23),

$$\begin{aligned}\Delta V(k)=&\,x^{\mathrm{T}}(k)Qx(k)+\hat{M}^{\mathrm{T}}(x(k))Z\hat{M}(x(k))-x^{\mathrm{T}}(k)Px(k)-x^{\mathrm{T}}(k-h)Qx(k-h)-x^{\mathrm{T}}(k)Zx(k)\\&+\left\|\tilde{w}^{\mathrm{T}}(k+1)\right\|^2-\left\|\tilde{w}^{\mathrm{T}}(k)\right\|^2+Z(k).\end{aligned} \tag{24}$$

where

$$\begin{aligned}Z(k)=&x^{\mathrm{T}}(k)A^{\mathrm{T}}P\tilde{g}(x(k))u(k)+x^{\mathrm{T}}(k-h)A_d^{\mathrm{T}}P\tilde{g}(x(k))u(k)+u^{\mathrm{T}}(k)\hat{g}^{\mathrm{T}}(x(k))P\tilde{g}(x(k))\\&+u^{\mathrm{T}}(k)\tilde{g}^{\mathrm{T}}(x(k))PAx(k)+u^{\mathrm{T}}(k)\tilde{g}^{\mathrm{T}}(x(k))PA_d^{\mathrm{T}}x(k-h)+u^{\mathrm{T}}(k)\tilde{g}^{\mathrm{T}}(x(k))P\hat{g}(x(k))u(k)\\&+u^{\mathrm{T}}(k)\tilde{g}^{\mathrm{T}}(x(k))P\tilde{g}(x(k))u(k)-x^{\mathrm{T}}(k)A^{\mathrm{T}}Z\tilde{g}(x(k))u(k)-x^{\mathrm{T}}(k-h)A_d^{\mathrm{T}}Z\tilde{g}(x(k))u(k)\\&-u^{\mathrm{T}}(k)\hat{g}^{\mathrm{T}}(x(k))Z\tilde{g}(x(k))-u^{\mathrm{T}}(k)\tilde{g}^{\mathrm{T}}(x(k))ZAx(k)-u^{\mathrm{T}}(k)\tilde{g}^{\mathrm{T}}(x(k))ZA_d^{\mathrm{T}}x(k-h)\\&-u^{\mathrm{T}}(k)\tilde{g}^{\mathrm{T}}(x(k))Z\hat{g}(x(k))u(k)-u^{\mathrm{T}}(k)\tilde{g}^{\mathrm{T}}(x(k))Z\tilde{g}(x(k))u(k)+u^{\mathrm{T}}(k)\tilde{g}^{\mathrm{T}}(x(k))Zx(k)\\&+x^{\mathrm{T}}(k)Z\tilde{g}(x(k))u(k).\end{aligned} \tag{25}$$

Collecting the terms together and substitute control law Equation (5) in Equation (24) yields

$$\begin{aligned}\Delta V(k)=&-\|\hat{w}(k+1)\|^2+\left[(\hat{w}(k))+\left(x^{\mathrm{T}}(k)(Q-P-Z)x(k)\right)^{1/2}+\left(\hat{M}^{\mathrm{T}}(x(k))Z\hat{M}(x(k))\right)^{1/2}\right]^2\\&+S(k)+Z(k)\end{aligned} \tag{26}$$

where

$$\begin{aligned}S(k)=&-2(\hat{w}(k))\left(\hat{M}^{\mathrm{T}}(x(k))Z\hat{M}(x(k))\right)^{1/2}-2(\hat{w}(k))\left(x^{\mathrm{T}}(k)(Q-P-Z)x(k)\right)^{1/2}\\&-2\left(x^{\mathrm{T}}(k)(Q-P-Z)x(k)\right)^{1/2}\left(\hat{M}^{\mathrm{T}}(x(k))Z\hat{M}(x(k))\right)^{1/2}-x^{\mathrm{T}}(k-h)Qx(k-h).\end{aligned} \tag{27}$$

with tuning law in Equation (8) , Equation (26) will be,

$$\Delta V(k)=S(k)+Z(k) \tag{28}$$

substituting Equations (25), (27) in Equation (28)

$$\begin{aligned}\Delta V(k) = &-2\left(\hat{w}(k)\right)\left(\hat{M}^{\mathrm{T}}\left(x(k)\right)Z\hat{M}\left(x(k)\right)\right)^{1/2} - 2\left(\hat{w}(k)\right)\left(x^{\mathrm{T}}(k)(Q-P-Z)x(k)\right)^{1/2}\\ &-2\left(x^{\mathrm{T}}(k)(Q-P-Z)x(k)\right)^{1/2}\left(\hat{M}^{\mathrm{T}}\left(x(k)\right)Z\hat{M}\left(x(k)\right)\right)^{1/2}\\ &+x^{\mathrm{T}}(k)\left[A^{\mathrm{T}}G^{\mathrm{T}}PGA - A^{\mathrm{T}}G^{\mathrm{T}}ZGA - A^{\mathrm{T}}G^{\mathrm{T}}Z - ZGA\right]x(k)\\ &+x^{\mathrm{T}}(k)\left[A^{\mathrm{T}}G^{\mathrm{T}}PGA_d - A^{\mathrm{T}}G^{\mathrm{T}}ZGA_d - ZGA_d\right]x(k-h)\\ &+x^{\mathrm{T}}(k-h)\left[A_d^{\mathrm{T}}G^{\mathrm{T}}PGA - A_d^{\mathrm{T}}G^{\mathrm{T}}ZGA - A_d^{\mathrm{T}}ZG^{\mathrm{T}}\right]x(k)\\ &+x^{\mathrm{T}}(k-h)\ \left[A_d^{\mathrm{T}}G^{\mathrm{T}}PGA_d - A_d^{\mathrm{T}}G^{\mathrm{T}}ZGA_d - Q\right]x(k-h).\end{aligned} \tag{29}$$

Manipulating the nonquadratic terms using the following inequality $\sqrt{ab} \le \frac{a+b}{2}$ (which turns into equality if and only if $a = b$ we get,

$$\begin{aligned}\Delta V(k) \le &-4\ \left(\left(\hat{w}(k)\right)\left(\hat{M}^{\mathrm{T}}\left(x(k)\right)Z\hat{M}\left(x(k)\right)\right)^{1/2}\left(x^{\mathrm{T}}(k)(Q-P-Z)x(k)\right)^{1/2}\right)^{1/2}\\ &-2\left(x^{\mathrm{T}}(k)(Q-P-Z)x(k)\right)^{1/2}\left(\hat{M}^{\mathrm{T}}\left(x(k)\right)Z\hat{M}\left(x(k)\right)\right)^{1/2}\\ &+x^{\mathrm{T}}(k)\left[A^{\mathrm{T}}G^{\mathrm{T}}PGA - A^{\mathrm{T}}G^{\mathrm{T}}ZGA - A^{\mathrm{T}}G^{\mathrm{T}}Z - ZGA\right]x(k)\\ &+x^{\mathrm{T}}(k)\left[A^{\mathrm{T}}G^{\mathrm{T}}PGA_d - A^{\mathrm{T}}G^{\mathrm{T}}ZGA_d - ZGA_d\right]x(k-h)\\ &+x^{\mathrm{T}}(k-h)\left[A_d^{\mathrm{T}}G^{\mathrm{T}}PGA - A_d^{\mathrm{T}}G^{\mathrm{T}}ZGA - A_d^{\mathrm{T}}ZG^{\mathrm{T}}\right]x(k)\\ &+x^{\mathrm{T}}(k-h)\left[A_d^{\mathrm{T}}G^{\mathrm{T}}PGA_d - A_d^{\mathrm{T}}G^{\mathrm{T}}ZGA_d - Q\right]x(k-h).\end{aligned} \tag{30}$$

where

$$a = -2\left(\hat{w}(k)\right)\left(\hat{M}^{\mathrm{T}}\left(x(k)\right)Z\hat{M}\left(x(k)\right)\right)^{1/2} \tag{31}$$

$$b = -2\left(\hat{w}(k)\right)\left(x^{\mathrm{T}}(k)(Q-P-Z)x(k)\right)^{1/2} \tag{32}$$

In Equation (30), $\Delta V(k)$ is guaranteed to remain negative as long as

$$\begin{bmatrix} A^{\mathrm{T}}G^{\mathrm{T}}PGA - A^{\mathrm{T}}G^{\mathrm{T}}zGA - A^{\mathrm{T}}G^{\mathrm{T}}z - zGA & A^{\mathrm{T}}G^{\mathrm{T}}PGA_d - A^{\mathrm{T}}G^{\mathrm{T}}zGA_d - GzA_d \\ * & A_d^{\mathrm{T}}G^{\mathrm{T}}PGA_d - A_d^{\mathrm{T}}G^{\mathrm{T}}zGA_d - Q \end{bmatrix} < 0$$

Since the first two terms in Equation (30) are satisfying the condition in Equation (9), (10), next four terms are satisfying the LMI in Equation (11). Therefore, we conclude that the system in Equation (4) is stable with control law Equation (5) and LMI in Equation (11).

6. SIMULATION RESULTS

In this section, a numerical example is presented to validate the performance and effectiveness of the nonlinear discrete-time system proposed in Equation (4). Consider the set of parameters for the given system

$$A=\begin{bmatrix} 0.8 & 0.4 \\ -0.14 & 0.1 \end{bmatrix},$$

and

$$g(x(k))=\begin{bmatrix} \dfrac{1.4x_1^2(k)}{1+x_1^2(k)} \\ \dfrac{x_1(k)}{1+x_1^2(k)+x_2^2(k)} \end{bmatrix}$$

The fixed time delay is assumed to be $h=2$. The proposed system has initial condition of states x_1 and x_2 are chosen as $[0.01 \quad 0,02]^T$. The LMI in H1) are solved by using Matlab LMI Toolbox and the values of P, Q, z and G are obtained as

$$P=\begin{bmatrix} 108.8411 & 0 \\ 0 & 108.8411 \end{bmatrix},$$

$$G=10^{-10}\begin{bmatrix} -0.1198 & 0.0001 \\ 0.0001 & 0.0247 \end{bmatrix},$$

The trajectories of the system states x_1 and x_2 are shown in Figure 2 and Figure 3. It is observed in Figure 2 and Figure 3 that the states of the sliding motion approach to zero quickly. Figure 4demonstrates that the controller robustly stabilizes the system by sliding mode technique with fixed time delay. The simulation results confirm the stability of the system and show the effectiveness of proposed scheme.

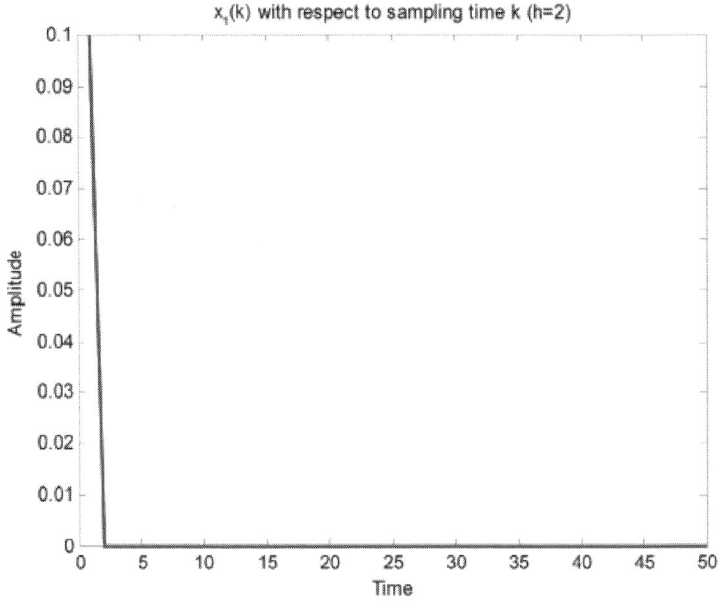

Figure 2. $x_1(k)$with respect to sample time $k \quad (h=2)$.

7. CONCLUSION

Vishal Goyal, Vinay Kumar Deolia, Tripti Nath Sharma In this paper, a sliding mode control for a class of unknown nonlinear discrete-time system is proposed ,which results in small chattering motion in both control signal and system output. A Chebyshev Neural Network is used to approximate the unknown system dynamics. A new learning algorithm for neural network approximation is proposed. This neural network based sliding mode control approach guarantees the system state trajectory to the defined sliding surface. An LMI based sufficient condition for the asymptotic stability of the sliding mode dynamics is derived by means of a Lyapunov-Krasovskii approach. Simulation results are validating the effectiveness of proposed scheme.

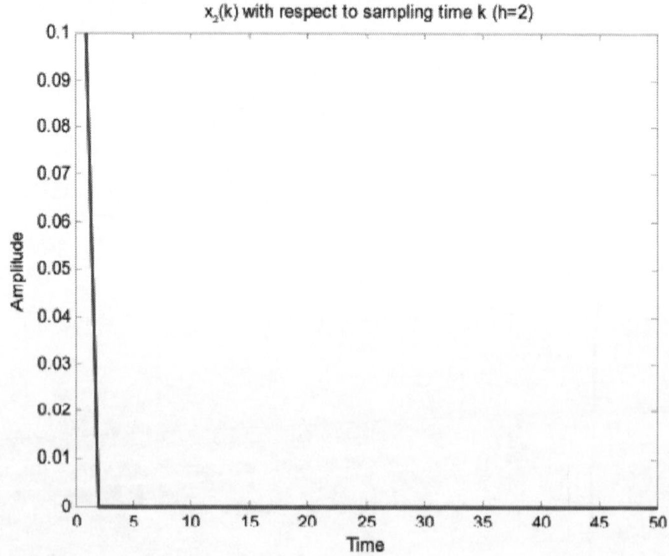

Figure 3. $x_2(k)$ with respect to sample time $k \quad (h=2)$.

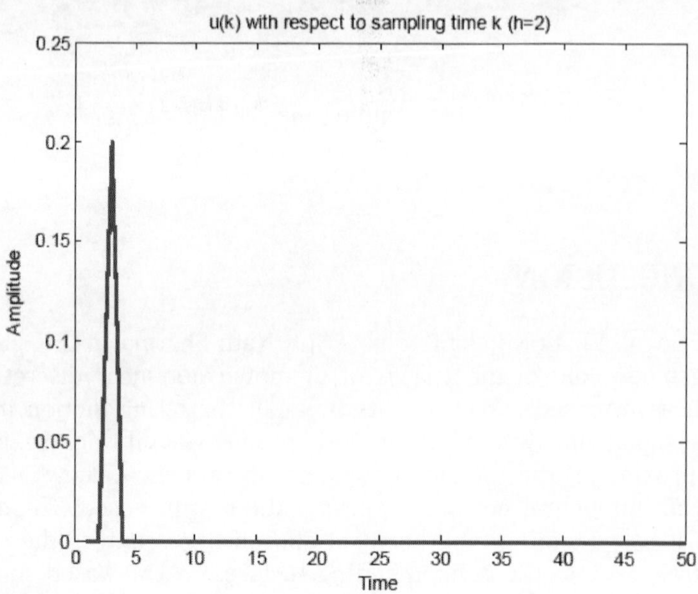

Figure 4. $u(k)$ with respect to sample time $k \quad (h=2)$.

REFERENCES

1. Utkin, V.I. (1977) Variable Structure Systems with Sliding Modes. IEEE Transactions on Automatic Control, AC-22, 212-222.

2. Utkin, V.I. (1978) Sliding Modes and Their Applications in Variable Structure Systems. Nauka, Moscow.

3. Sarpturk, S.Z., Istefanopolos, Y. and Kaynak, O. (1987) On the Stability of Discrete-Time Sliding Mode Control Systems. IEEE Transactions on Automatic Control, 32, 930-932.

4. Jeung, E.T., Oh, D.C., Kim, J.H. and Park, H.B. (1996) Robust Controller Design for Uncertain Systems with Time Delays, LMI Approach. Automatica, 32, 1229-1231.

5. Yue, D. (2004) Robust Stabilization of Uncertain Systems with Unknown Input Delay. Automatica, 40, 331-336

6. Basin, M.V., Perez, J., Acosta, P. and Fridman, L. (2006) Optimal Filtering for Nonlinear Polynomial Systems over Linear Observations with Delay. International Journal of Innovative Computing Information and Control, 2, 863-874.

7. Boukas, E.K. and Al-Muthairi, N.F. (2006) Delay-Dependent Stabilization of Singular Linear Systems with Delays. International Journal of Innovative Computing Information and Control, 2, 283-291.

8. Chen, M., Lam, J. and Xu, S. (2006) Memory State Feedback Guaranteed Cost Control for Neutral Delay Systems. International Journal of Innovative Computing Information and Control, 2, 293-303

9. Ge, S.S., Hong, F. and Lee, T.H. (2004) Adaptive Neural Control of Nonlinear Time-Delay Systems with Unknown Virtual Control Coefficients. IEEE Transactions on Systems, Man, and Cybernetics, Part B: Cybernetics, 34, 449-516.

10. Ge, S.S., Hong, F. and Lee, T.H. (2003) Adaptive Neural Network Control of Nonlinear Systems with Unknown Time Delays. IEEE Transactions on Automatic Control, 48, 2004-2010.

11. Xia, Y. and Jia, Y. (2003) Robust Sliding-Mode Control for Uncertain Time-Delay Systems: An LMI Approach. IEEE Transactions on Automatic Control, 48, 1086-1092.

12. Wang, Z., Huang, B. and Unbehauen, H. (1999) Robust Reliable Control for a Class of Uncertain Nonlinear State-De- layed System. Automatica, 35, 955-963.

13. Shyu, K. and Yan, J. (1993) Robust Stability of Uncertain Time-Delay Systems and Its Stabilization by Variable Structure Control. International Journal of Control, 57, 237-246.

14. Roh, Y.H. and Oh, J.H. (1999) Robust Stabilization of Uncertain Input-Delay Systems by Sliding Mode Control with Delay Compensation. Automatica, 35, 1861-1865.

15. Basin, M.V., Gonzalez, J.R. and Fridman, L. (2003) Optimal and Robust Sliding Mode Control for Linear Systems with Time Delays in Control Input. International Journal of Pure and Applied Mathematics, 5, 395-420.

16. Xia, Y., Liu, G.P., Shi, P., Chen, J. and Rees, D. (2008) Robust Delay-Dependent Sliding Mode Control for Uncertain Time-Delay Systems. International Journal of Robust and Nonlinear Control, 18, 1142-1161.

17. Yan, M. and Shi, Y. (2008) Robust Discrete-Time Sliding Mode Control for Uncertain Systems with Time—Varying State Delay. IET Control Theory & Applications, 2, 662-674.

18. Zhang, T., Ge, S.S. and Hang, C.C. (1999) Design and Performance Analysis of a Direct Adaptive Controller for Nonlinear Systems. Automatica, 35, 1809-1817.

19. Chen, S. and Billings, S.A. (1992) Neural Networks for Nonlinear Dynamic System Modelling and Identification. International Journal of Control, 56, 319-346.

20. Jagannathan, S. and Lewis, F.L. (1996) Identification of Nonlinear Dynamical Systems Using Multilayered Neural Networks. Automatica, 36, 1707-1712.

21. Niu, Y., Lam, J., Wang, X. and Ho, D.W.C. (2003) Sliding-Mode Control for Nonlinear State-Delayed Systems Using Neural-Network

Approximation. IEEE Proceedings—Control Theory and Applications, 150, 233-239.

22. Munoz, D. and Sbarbaro, D. (2000) An Adaptive Sliding-Mode Controller for Discrete Nonlinear Systems. IEEE Transaction on Industrial Electronics, 47, 574-581.

23. de Jesús Rubio, J. and Yu, W. (2006) Discrete-Time Sliding-Mode Control Based on Neural-Networks. Springer-Verlag, Berlin, 956-961.

24. Patra, J.C. and Kot, A.C. (2002) Nonlinear Dynamic System Identification Using Chebyshev Functional Link Artificial Neural Networks. IEEE Transaction on Systems, Man and Cyberntics, 32, 505-511.

25. Purwar, S., Kar, I.N. and Jha, A.N. (2007) Nonlinear System Identification Using Neural Networks. IETE Journal of Research, 53, 35-42.

26. Deolia, V.K., Purwar, S. and Sharma, T.N. (2012) Stabilization of Unknown Nonlinear Discrete-Time Delay Systems Based on Neural Network. Intelligent Control and Automation, 3, 337-345.

27. Namatame, A. and Ueda, N. (1992) Pattern Classification with Chebyshev Neural Network. International Journal Neu- ral Network, 3, 23-31.

28. Lee, T.T. and Jeng, J.T. (1998) The Chebyshev Polynomial Based Unified Model Neural Networks for Functions Approximations. IEEE Transactions on Systems, Man & Cybernetics, Part B, 28, 925-935.

29. Purwar, S., Kar, I.N. and Jha, A.N. (2008) Adaptive Output Feedback Tracking Control of Robot Manipulators Using Position Measurements Only. Expert Systems with Applications, 34, 2789-2798.

30. Purwar, S., Kar, I.N. and Jha, A.N. (2005) On-Line System Identification of Complex Systems Using Chebyshev Neural Networks. Applied Soft Computing, 7, 364-372.

9

PID STABILIZATION OF LINEAR NEUTRAL TIME-DELAY SYSTEMS IN A NUMERICAL APPROACH

Hassan Farokhi Moghadam, Nastaran Vasegh, Seyed Zeinolabedin Moussavi

Faculty of Electrical & Computer Engineering, Shahid Rajaee Teacher Training University, Tehran, Iran

ABSTRACT

In this paper, the stabilization of neutral time-delay systems is investigated. An efficient numerical approach is presented in an algorithm to establish results so that stability of such systems is achieved and stabilizing PID parameters are determined directly. It is based on determining the rightmost characteristic roots and Nyquist plot. The NewtonRaphson's iterative method based on Lambert W function is used for the calculation of these stabilizing roots directly from the closed-loop characteristic equation of the neutral time-delay system and then stability is checked by Nyquist plot and step response of closed-loop system. Two numerical examples are included to illustrate the effectiveness of the proposed approach.

KEYWORDS

Neutral Time-Delay Systems; PID Controller; Stability; Iterative Method; Rightmost Characteristic Root; Nyquist Plot

1. INTRODUCTION

Time-delay or hereditary systems are also called systems with aftereffect or dead-time [1]. Time-delay is usually unavoidable in many mechanical and electrical systems. It often appears in control and real-world engineering systems. The presence of delays complicates the system analysis and the control design [2]. In some cases, the presence of a small delay may destabilize the system. Because of the destabilizing nature of the delayed states in a system, stability analysis of time-delay systems becomes an important area of research (see [3-7] and references therein).

As mentioned in [8], time-delay system of neutral type, where the poles lie in a band centered on the imaginary axis is the most delicate case, and the object of our present study.

In a neutral time-delay system, the time-derivative of the state depends on both the current and delayed stated and also the past derivative [9]. Stability of this type of system is playing an increasingly important role in control engineering and has received considerable attention and also has been studied extensively in the literature (see [10-14] and references therein).

PID stabilization of this type of system is the main consideration in this paper. PID (Proportional-IntegralDerivative) controller design was done by Ziegler and Nichols [15] for the first time and it is still very applicable and efficient solution to many real-world control problems with its relatively simple way of tuning. According to a survey paper [16], more than 90% of controllers are of PID structure; even complicated control techniques also embed PID algorithms [17].

It is the most common control law for SISO systems in control engineering [18]. For this controller many design methodologies have been presented. For example, in [19], the proposed method is based on decomposing the numerator and denominator of the plant transfer function into their even and odd parts and then computing the stabilizing values of the parameters of the controller for a given time-delay system. In [20-22], a mathematical generalization of the Hermite-Biehler theorem to find all stabilizing PID controllers for systems with time-delay has been used.

In this paper, we present an efficient approach for determining the stabilizing PID controller for neutral timedelay systems. It's worthy to note that Nyquist plot and step response of closed-loop system are also used to show the correctness of our proposed stabilizing method.

The organization of this paper is as follows: In Section 2, the main problem and the stabilizing algorithm are stated. Section 3 shows stabilization of linear neutral delayed systems. In two sub-sections we introduce rightmost characteristic root or stability determining characteristic roots and Nyquist plot which are of great role in this paper. Section 4 shows illustrating of the approach by two examples. Both MAPLE and MATLAB, the two popular mathematical softwares, are used for the programming. Simulation results will show the advantages of the approach. Finally, in Section 5, the main conclusions are summarized.

2. PROBLEM STATEMENT

PID is a combination of three controllers: proportional, integral and derivative controller. Thus, the PID controller can be understood as a controller that takes the present, the past, and the future of the error into consideration. The main transfer function for PID here in the sdomain is as follows:

$$K(s) = k_p + \frac{k_i}{s} + k_d s = k_1 + \frac{k_2}{s} + k_3 s. \tag{1}$$

which both equivalent forms are used in this paper. We consider the neutral time-delay systems with following general transfer function form in the s-domain:

$$G(s) = \frac{1}{p(s) + q(s)e^{-s\tau}} \tag{2}$$

where $\tau > 0$ is time-delay and p, q are real polynomials. According to [8], the system is of neutral type if degrees of p and q are the same.

PID stabilization of this type of system is the main consideration in this paper. The stabilization is based on an algorithm which is proposed as follows.

Proposed Algorithm

At first, it's worthy to note that the algorithm is based on the known continues pole placement at [23] which is stated in following steps:

1) Initialize PID controller parameters.

2) Compute the rightmost characteristic roots for closed-loop characteristic equation of the neutral-time delay system.

3) Stop when the rightmost characteristic roots are in the left half plane and assure stability of neutral timedelay system. Then the values of PID controller parameters are chosen as stabilizing parameters. In the other case, go to the next step.

4) Compute the inverse of sensitivity matrix (S_m) of the rightmost characteristic roots with respect to the PID controller parameters.

It's worthy to note that S_m is similar to what defined at [20] and obtained by

$$S_m = \left[s_{i,j} \right] \in R^{3\times3}, s_{i,j} = \frac{\partial \lambda_i}{\partial k_j} \tag{3}$$

where λ_i represents the rightmost characteristic roots.

5) Move the rightmost characteristic roots in the direction of the left half plane by applying a small change to the PID controller parameters as:

$$\Delta(k_1, k_2, k_3) = S_m \, \delta \tag{4}$$

where δ is a small change chosen optionally and it's better to be a matrix with small negative values because we want to shift the rightmost characteristic roots in the direction of the left half. It is considered here as $\delta = [-0.1 \; -0.1 \; -0.1]^T$.

6) Monitor the PID controller parameters. New PID controller parameters should be determined by adding initial values of PID controller to the values which have been obtained at step d. It becomes:

$$K = (k_1, k_2, k_3) + \Delta(k_1, k_2, k_3) \tag{5}$$

Now go to step b.

It's worthy to note that K represents main stabilizing parameters of PID controller.

With respect to (1), we can rewrite (4) as

$$\begin{bmatrix} k_p \\ k_i \\ k_d \end{bmatrix} = \begin{bmatrix} \frac{\partial \lambda_1}{\partial k_p} & \frac{\partial \lambda_1}{\partial k_i} & \frac{\partial \lambda_1}{\partial k_d} \\ \frac{\partial \lambda_2}{\partial k_p} & \frac{\partial \lambda_2}{\partial k_i} & \frac{\partial \lambda_2}{\partial k_d} \\ \frac{\partial \lambda_3}{\partial k_p} & \frac{\partial \lambda_3}{\partial k_i} & \frac{\partial \lambda_3}{\partial k_d} \end{bmatrix} \begin{bmatrix} \delta_1 \\ \delta_2 \\ \delta_3 \end{bmatrix} \tag{6}$$

Advantages of this approach are shown by two illustrative examples in continuation.

3. STABILIZATION OF LINEAR NEUTRAL DELAYED SYSTEMS

In this section, we consider the problem of designing the stabilizing PID controller for the system with the transfer function in (2). The main goal of design is to find the values of PID controller parameters such that the closed- loop characteristic equation of the system is stable.

The stabilizing technique used in this paper is based on determining the rightmost characteristic roots and Nyquist plot. These two important basic principles are explained in brevity.

3.1. The Rightmost Characteristic Roots

Investigating the rightmost characteristic roots can be ensuring factor for stability analysis. Its importance can be seen more clearly when we know that a neutral time-delay system usually has an infinite number of roots, and it's very difficult to find out all the roots. It is usually required to find out the rightmost characteristic roots numerically [24]. In this paper the well-known Newton-Raphson's iterative method based on Lambert W function is used to determine the rightmost characteristic root.

In this paper we just define the following principles from [24,25] which are used in determining the rightmost characteristic root.

$$z = we^{w}, w, z \in C \tag{7}$$

where w = W(z) is the Lambert W function. According to [24], the solution W(z) has as many as infinite branches denoted by $W_i(z)$, i = 0, ±1, ±2, ···, $W_0(z)$ is the unique branch that is analytic at the origin $z=0$ and called the principal branch and it is used in computation of rightmost characteristic roots in this paper. It can be presented as below

$$W_0(z) = \sum_{n=1}^{\infty} \frac{(-n)^{n-1}}{n!} z^n \tag{8}$$

For fixed constants $a > 0, b$ it is defined that

$$F_i(\lambda) = a\lambda + b - W_i\left(\left(a\lambda + b - \Delta(\lambda)\right)e^{a\lambda+b}\right), \quad i = 0, \pm 1, \pm 2, \cdots \tag{9}$$

where $\Delta(\lambda)$ is the characteristic equation and $F_i(\lambda)$ is the ith branch of Lambert W function. Note that, rightmost characteristic root is obtained from $F_0(\lambda)=0$ which is based on the main branch defined in (8). Now using Newton-Raphson's iteration method results in: [24]

$$\lambda_{i+1} = \lambda_i - \frac{F(\lambda_i)}{F'(\lambda_i)}, \quad i = 0, 1, 2, \cdots \tag{10}$$

The iteration is stopped at step i if

$$|\lambda_i - \lambda_{i-1}| < \varepsilon \tag{11}$$

for a given small ε.

Therefore λ_i obtained from (10) is the rightmost characteristic root. To this end, initial guess (λ_0) is chosen optionally and freely which is a complex number. In this paper initial guess and time-delay are $\lambda_0 = 0.2 + 3j$ and $\tau = 1$, respectively.

It's worthy to note that the imaginary parts of the rightmost characteristic roots are not considered in calculations.

3.2. Nyquist Plot

At first we define the following theorem.

Theorem 1 [26]. A linear dynamic delay system is asymptotically stable if and only if the Nyquist diagram of

$$W(\lambda) = \frac{\Delta(\lambda)}{(\lambda+1)^n} \tag{12}$$

does not encircle the origin of the complex plane.

Where $\Delta(\lambda)$ is the characteristic equation and n is the degree of it.

Stability is guaranteed in this paper, if Nyquist plot of

$$W(j\omega) = \frac{\Delta(j\omega + \text{Re}\,\lambda_i + \mu)}{(\lambda+1)^n} \tag{13}$$

does not encircle the origin of the complex plane for very small $\mu > 0$ [14]. In this paper μ is considered as 0.001.

The rightmost characteristic roots are tested by putting in this theorem. If the Nyquist plot does not encounter the origin, then we will claim that rightmost characteristic roots have been computed correctly and the stability is achieved. This approach is clarified in following illustrative examples.

4. NUMERICAL ILLUSTRATIVE EXAMPLES

To illustrate the usefulness of the proposed method, we present the following examples. Example 1 shows the application of the approach on an unstable transfer function with the second-order characteristic equation and Example 2 is a third-order time-delay system of neutral type.

4.1. Example 1

In this example we consider a following unstable transfer function neutral time-delay system

$$G(s)=\frac{1}{(s+1)(1-e^{-s})}$$

which has been considered in [8].

We will show that it can be stabilized (nominally) by a rational controller. The closed-loop characteristic equation can be rewritten as

$$\Delta(\lambda)=(k_d+1)\lambda^2+(k_p+1)\lambda+k_i-(\lambda^2+\lambda)e^{-\lambda} \tag{14}$$

In both examples, the initial value of PID controller parameters is chosen as:

$$K=\left[k_p\ k_i\ k_d\right]^T=[0.2\ 0.5\ 0.1]^T$$

It can be tested that the system is unstable yet and initial value should not result in stability directly.

By implicit differentiation, we have:

$$\frac{\partial\lambda_i}{\partial k_p}=\frac{-\lambda_i}{\alpha(\tau,\lambda_i)} \tag{15}$$

$$\frac{\partial\lambda_i}{\partial k_i}=\frac{-1}{\alpha(\tau,\lambda_i)} \tag{16}$$

$$\frac{\partial \lambda_i}{\partial k_d} = \frac{-\lambda_i^2}{\alpha(\tau, \lambda_i)} \tag{17}$$

where

$$\alpha(\tau, \lambda_i) = \tau e^{-\lambda_i \tau} \lambda_i^2 + \left(2k_d + 2 + \tau e^{-\lambda_i \tau} - 2e^{-\lambda_i \tau}\right)\lambda_i$$
$$+ k_p + 1 - e^{-\lambda_i \tau}$$

Now based on the step d of proposed algorithm, matrix of the PID controller parameters becomes:

$$\Delta(k_p, k_i, k_d) = [-0.1121\ -0.4766\ 0.514]^T$$

which cannot stabilize the system yet. So the algorithm is repeated. Based on the (5) and (6), we have:

$$\boldsymbol{K} = [1.9\ 0.015\ 0.614]^T. \tag{18}$$

Now rightmost characteristic roots are obtained as (imaginary parts have been dismissed): $\lambda_1 = -0.006718$, $\lambda_2 = -0.012223$, $\lambda_3 = -0.012232$.

It's great that all characteristic roots are in the left half plane and can stabilize the neutral system. Therefore values in (18) become the stabilizing PID parameters. **Figure 1** which is plotted based on (13) shows that Nyquist plot is in the right half plane and does not encounter the origin. According to theorem 1 the neutral system is stable. So from iteration number 2 on, the rightmost characteristic roots and consequently the stabilizing parameters, are obtained. The stabilization is also proved by the closed-loop step response which is shown in **Figure 2**.

Moreover, as shown in **Figure 3**, the curve of the real parts of the rightmost characteristic roots with respect to the delay can be produced numerically by means of the proposed algorithm, which is asymptotically stable for wide range of time delays and it is shown in Figure 3 just for $\tau \in [0, 5000]$.

4.2. Example 2

In this example we consider another neutral time-delay system with following transfer function

$$G(s)=\frac{1}{(s+1)(s+2+0.2se^{-s})}$$

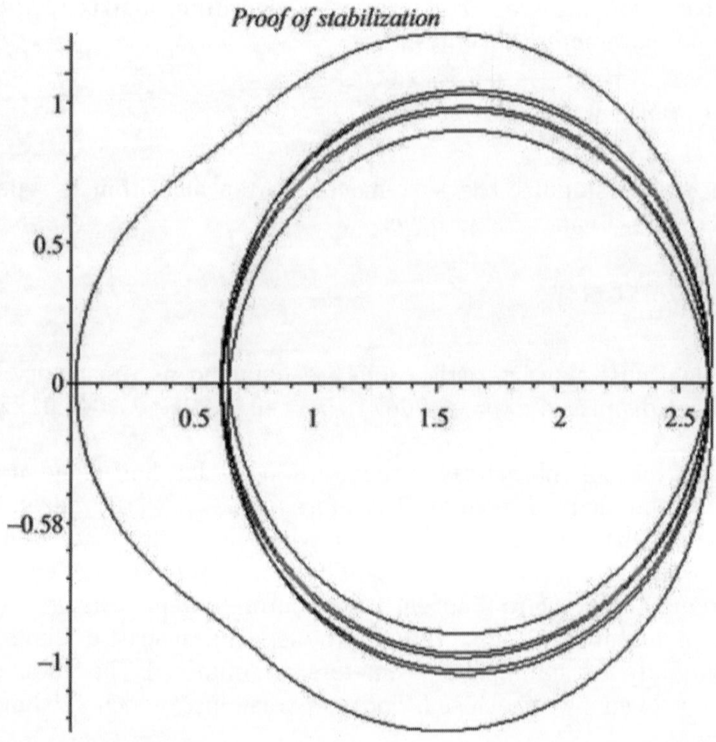

Figure 1. Proof of stabilization of Example 1 by Nyquist plot.

whose closed-loop characteristic equation of system can be given by

$$\Delta(\lambda)=(1+0.2e^{-\lambda})\lambda^3+(k_d+3+0.2e^{-\lambda})\lambda^2 +(k_p+2)\lambda+k_i \quad (19)$$

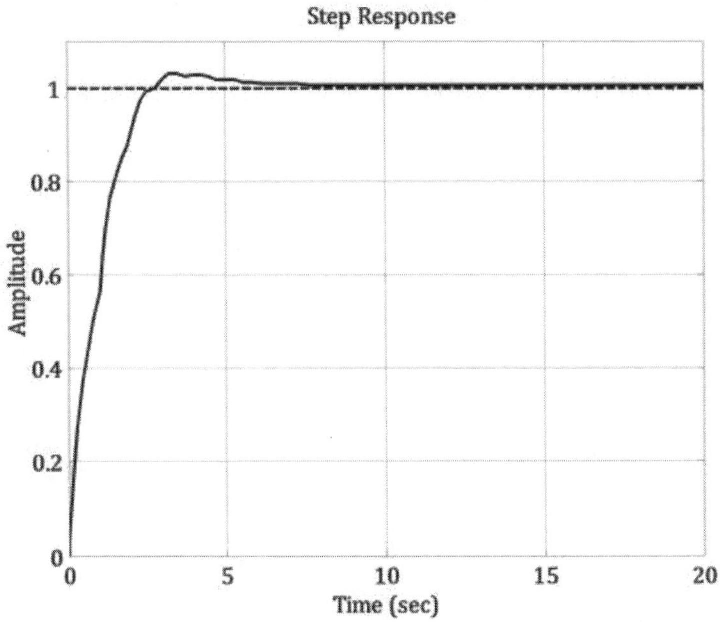

Figure 2. Step response of closed-loop system at Example 1 with PID parameters at (18).

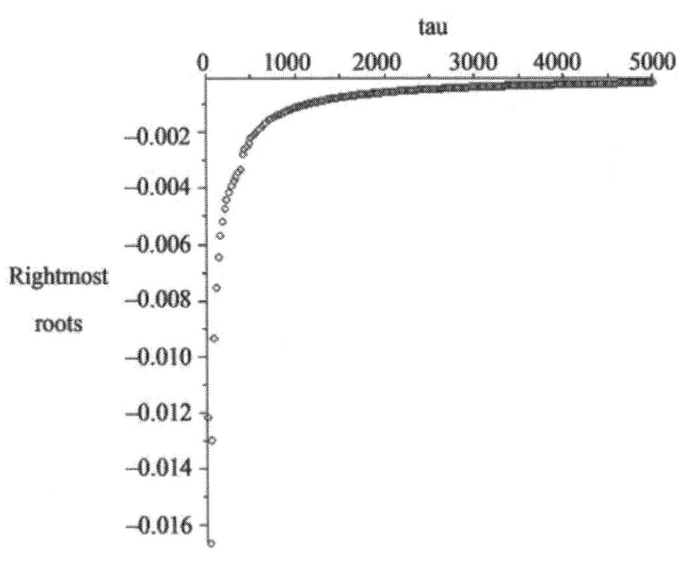

Figure 3. The stabilizing rightmost characteristic roots for wide range of time delays.

It can be tested that initial values have not been resulted in stability directly. By implicit differentiation, and the procedure like example 1, at iteration number 3 the rightmost characteristic roots and consequently the stabilizing parameters are obtained. These are obtained as $\lambda_1 = -0.2183, \lambda_2 = -0.3496, \lambda_3 = -0.3505$ and $K = [1.0904\ 0.8253\ 0.0701]^T$ respectively.

So procedure is stopped successfully. Nyquist plot in **Figure 4** shows the stability of neutral system for the first resultant characteristic root ($\lambda_1 = -0.2183$). The result is similar for the others and is dismissed here for brevity.

By continuing the algorithm, at the 5th iteration, $K = [1.61\ 1.59\ 0.01]^T$ and at the 8th iteration, $K = [2.2\ 2.37\ 0.0012]^T$ have been obtained, respectively. Closed-loop step response is shown in **Figure 5** for PID parameters at these different iterations.

It is worthy to note that this approach does not always stop successfully in the second or third stage. Sometime it is needed to be repeated more to get the stabilizing rightmost characteristic roots and guaranteed stability. If the desired stabilizing rightmost characteristic roots have not been obtained after iteration number 5, it is proposed to change the initial values of PID controller and start the algorithm from the beginning.

5. CONCLUSION

In the present paper, we presented an efficient and straightforward stabilizing PID controller design method for time-delay systems of neutral type which was free of mathematical complexities. Based on the proposed approach, stability was guaranteed if all the characteristic roots of closed-loop system had negative real parts. For determining the rightmost characteristic roots, the method was presented on the basis of Lambert W function and we managed to determine the stability directly. Delay value was chosen $\tau = 1$ for simplicity but the algorithm could produce a plot of the real part of the rightmost root with respect to the delay as shown in the first illustrative example. Numerical examples with time delay were presented for illustrating this approach. The results were satisfactory and stabilizing PID parameters were determined.

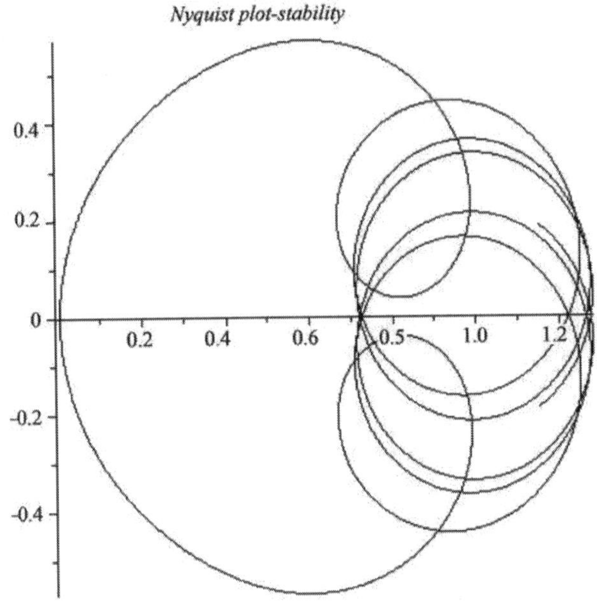

Figure 4. The Nyquist plot of Example 2 which does not encircle the origin.

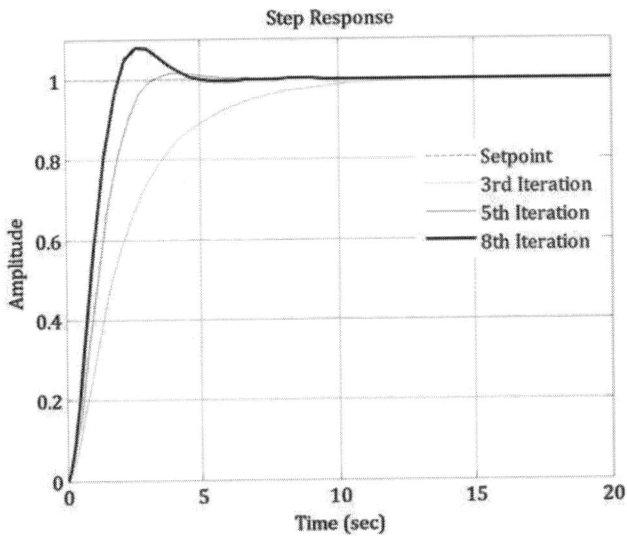

Figure 5. Step response of closed-loop system at Example 2 with PID parameters at different iterations.

REFERENCES

1. J. P. Richard, "Time-Delay Systems: An Overview of Some Recent Advances and Open Problems," Automatica, Vol. 39, No. 10, 2003, pp. 1667-1694.

2. J. E. Normey-Rico and E. F. Camacho, "Control of DeadTime Processes," Springer-Verlag, London, 2007.

3. V. B. Kolmanovskii and J. P. Richard, "Stability of Some Linear Systems with Delays," IEEE Transactions on Automatic Control, Vol. 44, No. 5, 1999, pp. 984-989.

4. J. K. Hale and S. M. V. Lunel, "Introduction to Functional Differential Equations," Springer-Verlag, New York, 1993.

5. L. Dugard and E. E. Verriest, "Stability and Control of Time-Delay Systems," Springer, New York, 1998.

6. S. I. Niculescu and R. Lozano, "On the Passivity of Linear Delay Systems," IEEE Transactions on Automatic Control, Vol. 46, No. 3, 2001, pp. 460-464.

7. K. Gu, V. L. Kharitonov and J. Chen, "Stability of TimeDelay Systems," Birkhauser, Boston, 2003.

8. J. R. Partingtona and C. Bonnet, "H∞ and BIBO Stabilization of Delay Systems of Neutral Type," Systems & Control Letters, Vol. 52, No. 3, 2004, pp. 283-288.

9. S. I. Niculescu, "Delay Effects on Stability: A Robust Control Approach," Springer, New York, 2001.

10. V. B. Kolmanovskii and A. D. Myshkis, "Applied Theory of Functional Differential Equations," Kluwer, Dordrecht, 1992.

11. V. B. Kolmanovskii and V. R. Nosov, "Stability of Functional Differential Equations," Academic Press, New York, 1986.

12. J. H. Park and O. Kwon, "On New Stability Criterion for Delay-Differential Systems of Neutral Type," Applied Mathematics and Computation, Vol. 162, No. 2, 2005, pp. 627-637.

13. V. Chellaboina, A. Kamath and W. M. Haddad, "TimeDomain Sufficient Conditions for Stability Analysis of Linear Neutral Time-Delay Systems," Proceedings of the 2007 American Control Conference, New York, 9-13 July 2007, pp. 4917-4918.

14. Z. H. Wang, "Numerical Stability Test of Neutral Delay Differential Equations," Hindawi Publishing Corporation, Cairo, 2008, pp. 1-10.

15. J. G. Ziegler and N. B. Nichols, "Optimum Settings for Automatic Controllers," Transactions on ASME, Vol. 64, 1942, pp. 759-768.

16. S. Yamamoto and I. Hashimoto, "Present Status and Future Needs: The View from Japanese Industry," Chemical Process Control—CPCIV: Proceedings of 4th International Conference on Chemical Process Control, Padre Island, 17-22 February 1991, pp. 1-28.

17. C. Dey and R. K. Mudi, "An Improved Auto-Tuning Scheme for PID Controllers," ISA Transactions, Vol. 48, No. 4, 2009, pp. 396-408.

18. B. Fang, "Computation of Stabilizing PID Gain Regions Based on the Inverse Nyquist Plot," Journal of Process Control, Vol. 20, No. 10, 2010, pp. 1183-1187.

19. N. Tan, "Computation of Stabilizing PI and PID Controllers for Processes with Time Delay," ISA Transactions, Vol. 44, No. 2, 2005, pp. 213-223.

20. K. W. Ho, A. Datta and S. P. Bhattacharya, "Generalizations of the Hermite-Biehler Theorem," Linear Algebra and Its Applications, Vol. 302-303, 1999, pp. 135-153.

21. K. W. Ho, A. Datta and S. P. Bhattacharya, "PID Stabilization of LTI Plants with Time-Delay," Proceedings of 42nd IEEE Conference on Decision and Control, Maui, 9-12 December 2003, pp. 4038-4043.

22. G. J. Silva, A. Datta and S. P. Bhattacharyya, "PID Controllers for Time-Delay Systems," Birkhäuser, Boston, 2005.

23. W. Michiels, K. Engelborghs, P. Vansevenant and D. Roose, "Continuous Pole Placement Method for Delay Equations," Automatica, Vol. 38, No. 5, 2002, pp. 747- 761.

24. Z. H. Wang and H. Y. Hu, "Calculation of the Rightmost Characteristic Root of Retarded Time-Delay Systems via Lambert W Function," Journal of Sound and Vibration, Vol. 318, No. 4-5, 2008, pp. 757-767.

25. R. M. Corless, G. H. Gonnet, D. E. G. Hare, D. J. Jeffrey and D. E. Knuth, "On the Lambert W function," Advances in Computational Mathematics, Vol. 5, No. 4, 1996, pp. 329-359.

26. H. Y. Hu and Z. H. Wang, "Dynamics of Controlled Mechanical Systems with Delayed Feedback," SpringerVerlag, Berlin Heidellberg, 2002.

INDEX